WORSHIP SPACE ACOUSTICS

Mendel Kleiner ◆ David Lloyd Klepper
Rendell R. Torres

ACOUSTICS: INFORMATION AND COMMUNICATION SERIES

Ning Xiang, Editor-in-Chief

Worship Space Acoustics
by Mendel Kleiner, David Lloyd Klepper, Rendell R. Torres

ISBN-13: 978-1-60427-037-2

Printed and bound in the U.S.A. Printed on acid-free paper.

10 9 8 7 6 5 4 3 2 1

Library of Congress Cataloging-in-Publication Data

Kleiner, Mendel, 1946–
Worship space acoustics / by Mendel Kleiner, David Lloyd Klepper, and Rendell R. Torres.
p. cm.
Includes index.
ISBN 978-1-60427-037-2 (hardcover : alk. paper)
1. Architectural acoustics. 2. Religious facilities. I. Klepper, David Lloyd, 1932- II. Torres, Rendell R., 1971- III. Title.
NA2800.K56 2010
690'.65—dc22
2010021236

Direct all inquiries to J. Ross Publishing, Inc., 5765 N. Andrews Way, Fort Lauderdale, FL 33309.

Phone: (954) 727-9333
Fax: (561) 892-0700
Web: www.jrosspub.com

Contents

About the Authors

Mendel Kleiner is professor of acoustics at Chalmers University of Technology, Gothenburg, Sweden, in charge of the Chalmers Room Acoustics Group since 1989. Dr. Kleiner obtained his Ph.D. in architectural acoustics in 1978. He was professor of architectural acoustics at Rensselaer Polytechnic Institute, Troy, New York from 2003 to 2005. Kleiner is responsible for teaching room acoustics, audio, electroacoustics, and ultrasonics in the Chalmers master program on sound and vibration (http://www.ta.chalmers.se/intro/index.php). Professor Kleiner has brought his Chalmers group to the international research front of predictive room acoustics calculation, audible simulation, and 3D sound. He returned to Chalmers in 2005 to continue leading the Chalmers Room Acoustics Group. He has over 50 publications, has presented more than 110 papers and keynote lectures, has led courses at international conferences on acoustics and noise control, and organized an international conference on acoustics. Kleiner's main research areas are computer simulation of room acoustics, electro-acoustic reverberation enhancement systems, room acoustics of auditoria, sound and vibration measurement technology, product sound quality, and psychoacoustics. He is the author of *Audio Technology and Acoustics* published by the Division of Applied Acoustics, Chalmers. Kleiner is a Fellow of the Acoustical Society of America, Chair for the Audio Engineering Society's Technical Committee on Acoustics and Sound Reinforcement and its Standards Committee on Acoustics.

David Lloyd Klepper, currently a student of Rabbinics at Yeshivat Beit Orot, Jerusalem, Israel, was formerly president of Klepper, Marshall, and King Acoustical Consultants and adjunct professor of architectural acoustics at City University, New York, New York. Previously, he was a senior consultant at Bolt, Beranek, and Newman with SM and SB degrees in electrical engineering from MIT. He served as an acoustical consultant for over 200 worship space buildings: the National Presbyterian Church, Washington, D.C.; St. Thomas Church Fifth Avenue, New York, New York; the Anglican Cathedral in Capetown, South Africa; River Road Baptist Church, Richmond, Virginia; Young Israel of Southfield, Michigan; and Boston's Holy Cross Cathedral. He is a pioneer in the application of digital delay and electronic simulation of reverberation in worship spaces and pew-back speech reinforcement. Klepper is the author of "Sound Systems for Reverberant Worship Spaces" that appeared in the *Journal of the Audio Engineering Society*, August 1970, two church acoustics papers, one concert hall paper, and a paper on safer audio entertainment in automobiles presented in *The Journal of the Acoustical Society of America.* Additionally, he authored "Acoustics of Assembly Places," *The Encyclopedia of Architecture,* coauthored with Mendel Kleiner on "Sound Amplification Systems," the *Encyclopedia of Science* and has to his credit 34 other published papers on acoustics, noise control, and electronic sound reinforcement systems. He served as the editor for *Sound Reinforcement Anthology I* (1978) and *Sound Reinforcement Anthology II* (1996), Audio Engineering Society, New York, New York. Klepper is a fellow of both the Audio Engineering Society and the Acoustical Society of America, and he is a member of the Institute of Noise Control Engineering and the American Guild of Organists. He is the recipient of the Silver Medal (Berliner Medal) from the AES.

Rendell R. Torres is a priest for the Roman Catholic Diocese of Albany, New York. Before the priesthood he was a tenure-track professor and director of the program in architectural acoustics at Rensselaer Polytechnic Institute (RPI) and continues to serve as an adjunct professor. He obtained his undergraduate degree in civil engineering from the University of California, Berkeley, his M.S. in engineering acoustics from Penn State University, State College, PA, and his Ph.D. in applied acoustics from Chalmers Tekniska Högskola (Chalmers University of Technology) in Gothenburg, Sweden. He pursued research in architectural acoustics and auralization with the Chalmers Room Acoustics Group in Sweden, at the Institute of Technical Acoustics in Aachen, Germany, and with the acoustics program at RPI. He has been invited to speak on his research for the Acoustical Society of America, the International Congress on Acoustics in Japan and Italy, and the Institute of Acoustics in the United Kingdom. He has published in the *Journal of the Acoustical Society of America* and in *Acustica*, the journal of the European Acoustics Association. He is also an active cellist.

This book has free material available for download from the Web Added Value™ resource center at *www.jrosspub.com*

At J. Ross Publishing we are committed to providing today's professional with practical, hands-on tools that enhance the learning experience and give readers an opportunity to apply what they have learned. That is why we offer free ancillary materials available for download on this book and all participating Web Added Value™ publications. These online resources may include interactive versions of material that appears in the book or supplemental templates, worksheets, models, plans, case studies, proposals, spreadsheets, and assessment tools, among other things. Whenever you see the WAV™ symbol in any of our publications, it means bonus materials accompany the book and are available from the Web Added Value™ Download Resource Center at www.jrosspub.com.

Downloads for *Worship Space Acoustics* include noise control calculations for office air supply systems, guidance for churches and non-Orthodox synagogues in choosing an organ, and an example in the use of Performance History Database (PHD) software.

PART I

INTRODUCTION

WORSHIP SPACE ACOUSTICS

And the LORD spoke unto Moses, saying, And let them make me a sanctuary, that I may dwell among them (Exodus 25:8).

Voice and music as the carriers of the divine message are elemental to worship. We hear God's voice speaking to us and the voices of other worshippers by both speech and song. Musical instruments are also a medium of the divine.

Thus, worship is the pinnacle of communication between God and man. Worship can be individual or corporate; however, communal worship usually takes place in a dedicated space. Such a worship space must be designed to assist the communication sought by worship.

Communication during worship can take various forms. Interior prayer and contemplation can be silent whereas other forms of worship such as prayer, song, music, and even dance will employ sound. To assist in effective communication through sound is, therefore, a requirement of any worship space. The communicative properties of a worship space are described by its acoustics. Room acoustics is the branch of acoustics—the science of sound—that relates to indoor sound. Although concert halls have been a primary subject of room-acoustics research, worship spaces are used more frequently and by more people, thus calling for a book of this nature.

To experience voice and music in an optimal way, a place for worship must be quiet and have suitable acoustics. For speech, the acoustical requirements differ from those for song or music. Music, both vocal and instrumental, has other requirements than does speech and the requirements must be met for both in nearly all worship spaces.

The acoustics of a room is affected not only by the way the desired sound travels or how the sound is distributed in the room but also by the presence of undesired sounds—noise. A room's acoustics can either add to or detract from the worship experience. The size, shape, and materials that make up a room all influence the way that sound is heard.

Sources of sound also can have different properties, requiring various types of acoustics for optimal functioning. The congregation, composed of many distinct individuals, will have varying requirements for a good listening experience based on language and age along with other characteristics.

This book is focused on worship spaces for the Abrahamic religions, that is, the monotheistic faiths of Judaism, Christianity, and Islam. The text provides an introduction to acoustics and sound reinforcement for those involved with worship spaces: architects and students of architectural acoustics, building consultants, contractors and suppliers, administrators, clergy, organists and organ builders, students and faculty of religious educational institutions, and laypersons with interests in religion and architecture. The book also offers comparisons and contrasts between facilities intended for different religious groups.

The book is divided into two parts. The first part discusses the methods and techniques of room acoustics optimization, that is, how the acoustics of large and small spaces are designed, implemented, and adjusted. It also discusses noise and its control and how acoustical privacy is attained. Finally, the first part describes sound reinforcement and other methods for improving clarity and reverberation by electro-acoustic means.

The second part of the book provides three chapters on the main types of Jewish, Christian, and Muslim worship spaces—synagogues, churches, and mosques. It reviews the characteristics of worship services pertinent to the various faiths and how these are provided for through acoustical design and sound reinforcement. It also surveys the historical background behind current practices, problems faced, and solutions found, in an attempt to broaden and deepen the understanding of these spaces for those involved in their design, construction, and use.

For those who teach and study architecture, acoustics, or religion, we hope this reference book provides the guidance for designing, constructing, and using worship spaces with optimal acoustics. Extensive notes that contain additional detailed information on a selection of the buildings used in the examples are provided.

1

FUNDAMENTALS—NATURE OF SOUND

1.1 SOME IMPORTANT PROPERTIES OF SOUND

Sounds are pressure oscillations in air created by vibrating surfaces and also as a result of turbulence. Fundamental, physically measurable properties of sound are frequency, amplitude, and wavelength. For our discussion, air is considered a continuous medium consisting of molecules that vibrate back and forth as they carry sound waves.

The rate of oscillation is called *frequency* (f) and is given in units of *hertz* (Hz). Frequency is determined by the number of times per second a small volume of air vibrates around its neutral position. A hertz is equal to one cycle of vibration per second. One complete vibration is movement of the volume from the neutral point to compression, back through the neutral point to rarefaction, and the return to the neutral position. The greater the number of vibrations or cycles per second, the higher the frequency. Sounds may be audible if the oscillation is within the frequency range of human hearing; that is, 20 Hz to 20 kHz. (1000 Hz equals a *kilohertz* [kHz]). Figure 1.1 shows how sound waves are generated by a vibrating tuning fork and illustrates that a pure tone is composed of rarefaction and compression.

Many people, particularly the elderly, are hearing impaired and that usually affects high-frequency reception more than low-frequency reception, therefore, they may not hear treble frequencies. Typically, many elderly people do not hear

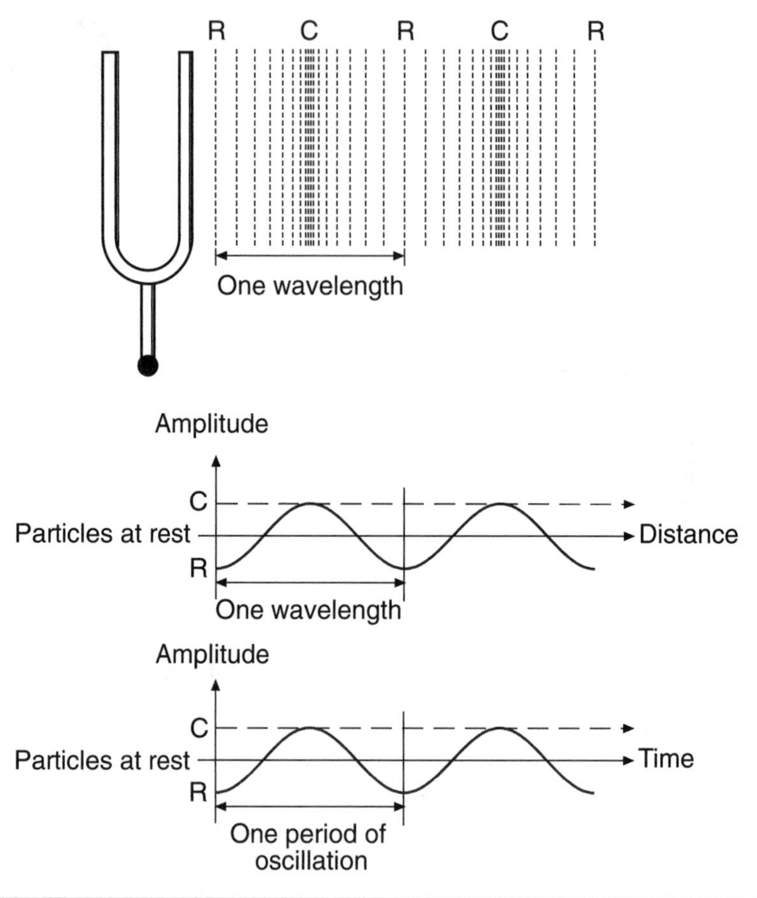

Figure 1.1 A vibrating tuning fork will cause sound waves consisting of rarefaction (R) and compression (C) of air. The undulating line in the two lower graphs shows the particle vibration (and sound pressure) as a function of distance and as a function of time.

sounds having frequencies above 4 to 6 kHz. For sounds to be audible, it is also necessary that they have sufficient pressure amplitude. Amplitude is determined by the number of molecules grouped together in compression (or rarefaction) and how far the molecules move back and forth. The denser the molecules and the greater displacement of their oscillation, the louder the sound.

The amplitude necessary for a sound to be heard varies widely over the range of human hearing. At extremely low or extremely high frequencies, the oscillations must be strong to be heard. It is easiest to hear sounds in the frequency range of 1000 to 4000 Hz (1 to 4 kHz). Sounds that are sufficiently strong induce pain in the ears. Sounds may also prevent the audibility of other sounds. This is called masking. A stronger sound will mask a weaker sound that has frequencies close to the frequency of the stronger sound.

Sound velocity is dependent on temperature and atmospheric pressure, but independent of amplitude and frequency. In air, sounds move slowly; at normal atmospheric pressure and at typical room temperature (20°C/68°F), the velocity of sound is 344 meters per second (m/s) (1,129 feet per second [f/s]). Because the velocity is so low, the wavelength of sound will be relatively large. The wavelength is the distance in the direction of sound travel between neutral points of adjacent compressions or rarefactions. Wavelength varies inversely with frequency as wavelength = sound velocity/frequency.

A tone with a frequency of 1 kHz has a wavelength of approximately 0.34 m (slightly more than 1 ft). The range of wavelength for audible sounds is 17 m to 1.7 cm (ca 19 ft to 5/8 in) corresponding to frequencies of 20 Hz and 20 kHz respectively. Figure 1.2 illustrates the relationship between frequency and wavelength.

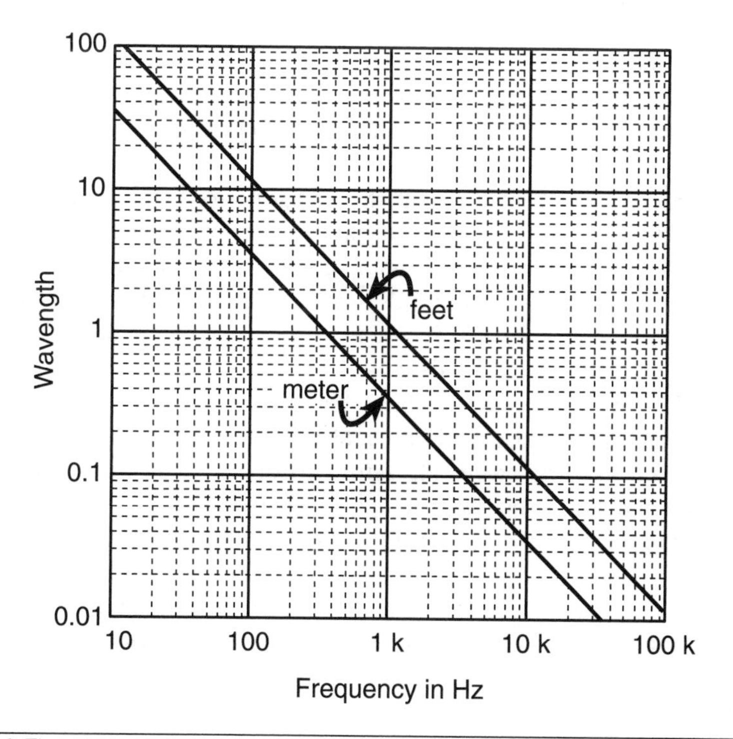

Figure 1.2 The relationship between wavelength and frequency for a sound velocity of 344 m/s (1,129 ft/s).

1.2 SOUND PRESSURE, SOUND PRESSURE LEVEL, AND SOUND LEVEL

Whereas one can describe the strength of a sound using the sound pressure or vibration amplitude in the wave, it is usually preferred to describe it using its sound pressure level (SPL). Sound pressure is expressed in the unit *Pascal*, which is equivalent to force per unit area; that is, Newton per square meter (N/m^2). SPL is expressed in decibels (dB).

The decibel scale, for our purposes, extends from 0 dB—the threshold of audibility with normal hearing—to approximately 130 dB—the threshold of pain. In an extremely quiet environment, the average person barely hears a mid-frequency sound level of 0 dB, and a 120 dB sound level just begins to tickle peoples' ears. Some rocket engines develop sound pressures that can be deadly to humans and can cause structural damage to buildings.

The decibel scale is used to express wide range, physically measurable quantities in a convenient manner. The typical measuring equipment measures sound pressure and expresses the measurement result as SPL on a decibel magnitude scale. The sound pressure scale within the range of human hearing extends over a million units and is difficult to handle, but the typical SPL range used is only 120 units and is much more convenient. There is sufficient reason to use the decibel scale when considering how people hear sound.

In an extremely quiet environment, the average person barely hears a 1 kHz tone at a SPL of 0 dB. The threshold of hearing at a frequency of 1 kHz is thus 0 dB.

The relative change in sound pressure corresponds to an increase in decibel, much the same way that it corresponds to an increase in loudness to our hearing. At a frequency of 1 kHz and at SPLs of 40 dB or more, a doubling of the subjectively perceived loudness of sounds typically corresponds to a change of approximately 10 dB. The scarcely noticeable difference in SPL is approximately 1 dB, which explains some of the popularity of SPL as a metric. The SPL scale can thus be viewed as a *temperature scale*. The sensitivity of hearing is greatest in the range of 1 to 4 kHz and diminishes toward high or low frequencies.

Because of the variation of hearing sensitivity with frequency, the sound pressure is usually measured using equipment that takes this variation into account; such measurements are given in *decibel A* (dBA) and are called the sound level in contrast to the SPL mentioned earlier. But it is frequently written as SPL_A or L_{pA}, both in units of dBA. Figure 1.3 shows some typical sound levels for various sound sources.

The other system for expressing the strength is sound intensity level (L_I or IL). The intensity level system is worth mentioning, but one cannot directly measure sound intensity (I). SPL, however, is easily measured. Practically, in most architectural acoustics situations, IL is equal to SPL, and 60 dB IL = 60 dB SPL.

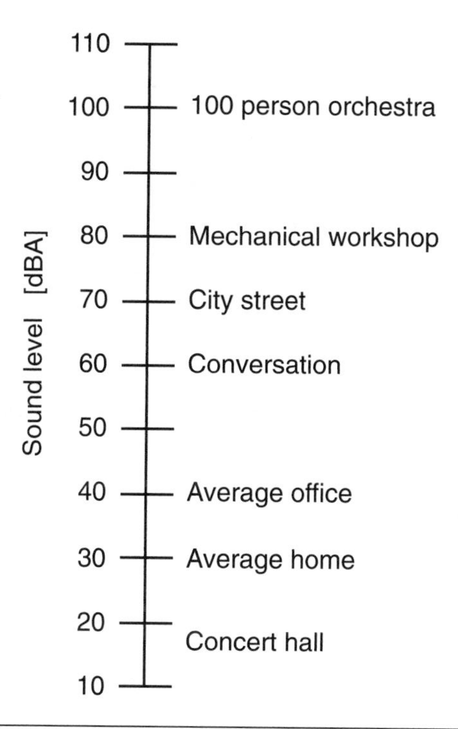

Figure 1.3 Approximate sound levels of some common sounds.

1.3 SOUND PRESSURE LEVEL AND SOUND POWER

Newton's second law of motion states that force equals mass × acceleration (F = ma). Work is done when a force is applied over a certain distance (equals force × distance) and is expressed in watts.

Power (P) in watts is a basic quantity of electrical, acoustical, or other energy per second. Electrical and acoustical watts are identical quantities, but the various forms provoke different responses in humans. Few watts are required to produce a (relatively) high level of sound because ears are sensitive. The ear can perceive sound corresponding to air movement the size of an oxygen molecule. If ten million people spoke simultaneously, they would be extremely noisy, but would, together, develop only 100 watts of power—enough to light a typical incandescent lightbulb.

Let us assume that we place a sound source in midair as shown in Figure 1.4, that it radiates sound waves spherically (meaning it is omnidirectional or radiates equally well in all directions), and that it produces a certain power (P). The sound power is spread over a sphere of increasing area as the sound wave radiates further from the source. Over any sphere at any distance from the source, the total sound

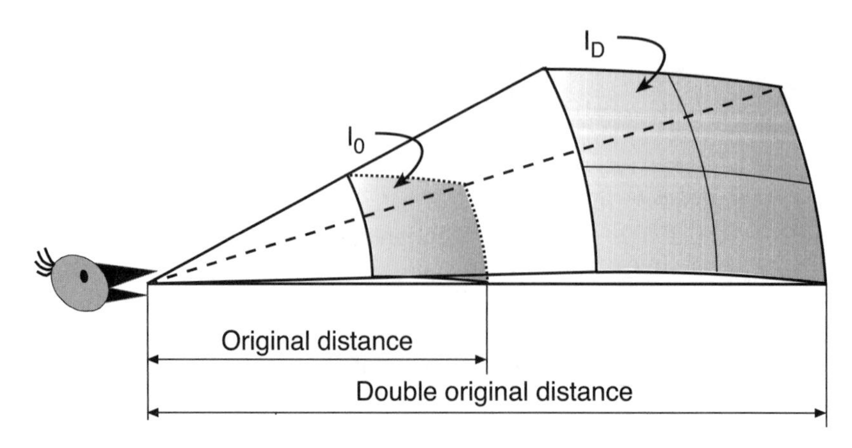

Figure 1.4 The sound power is spread over a sphere of increasing area as the sound wave radiates further from the source. For a small sound source that sends out spherical waves the sound intensity (power per unit area) is reduced to one-fourth as the distance is doubled, that is, $I_D = I_0/4$. This corresponds to a reduction of SPL by 6 dB. This is sometimes called the distance or geometrical attenuation.

power is the same, but the sound power per unit area decreases with increasing distance from the source. Typically, we are not interested in total sound power, but rather in the power per unit area (sound intensity) near our point in space because this is what affects the SPL or loudness at our ears.

As sound power is spread over a sphere of increasing area, so is the force of the moving air molecules associated with that power. Force per unit area is pressure (expressed in Pascal or N/m^2) and the pressure decreases as does the square root of the sound intensity, since I is proportional to the pressure $\times$ the velocity the pressure produces.

1.4 USING THE DECIBEL SCALE

The amplitude or strength of sound can be expressed directly as sound pressure. The pressure of the faintest sound we hear is 0.00002 Pa and the loudest is approximately 200 Pa. This is a pressure range of 10 million units, which is inconvenient for intuitive use in architectural engineering. Thus, the SPL is given in units of dBs.

The decibel scale is used to convert this large range to one that is more convenient:

a. The pressure of the faintest sound is defined as the reference pressure, $p_o = 0.00002$ Pa.

b. Divide the pressure of interest by the reference pressure, p/p_o.
c. The logarithm of this ratio is multiplied by a convenient (for our purposes) constant. We use $L_p = 20 \log(p/p_o)$.

This is the SPL expressed in dB. A useful scale is now only 120 units instead of a million units and is far more convenient. The relationship between pressure and SPL is presented graphically in Figure 1.5.

Applying the Inverse Square Rule to study the drop in SPL as a function of distance for the case shown in Figure 1.4, we find that the SPL drops with distance as depicted in Figure 1.6. If all frequencies are radiated the same way by the sound source the sound level and SPL will drop by the same number of dBs.

The SPL generally increases when sounds are added. Assume, for example, that at some point there is sound from two sound sources, each giving an SPL of 45 dB. What is the combined level of the two sources? Since the intensities increase, one must use the following scheme to obtain the combined level. (We assume that the sources, for example two loudspeakers, are not fed the same signal. Such cases must be handled by other methods.)

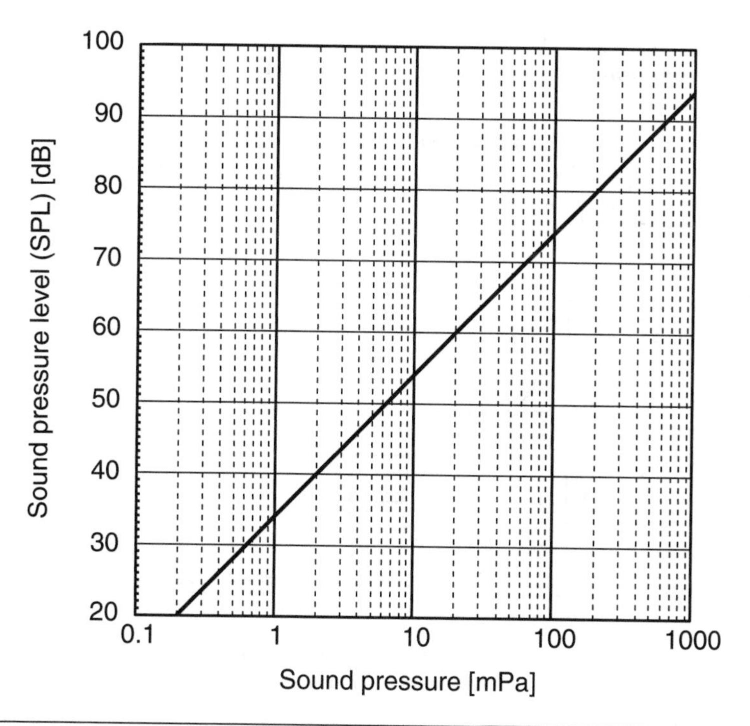

Figure 1.5 The relationship between sound pressure (in milliPascal [mPa]) and SPL in dBs.

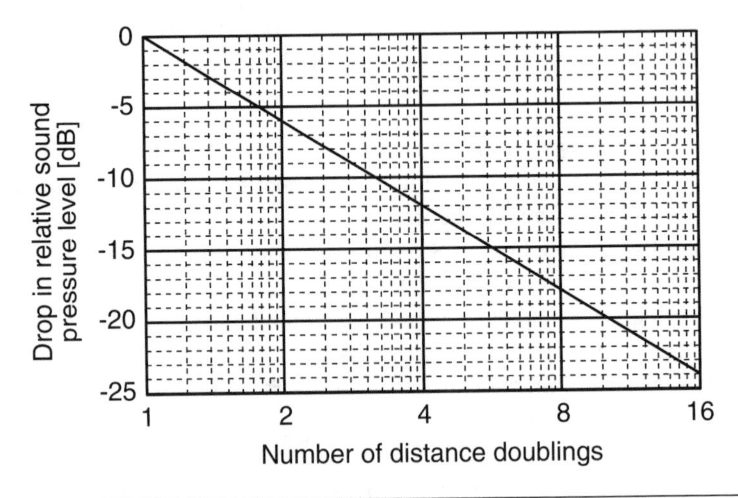

Figure 1.6 The distance attenuation for a spherical sound wave as a function of distance. Each distance doubling results in a SPL reduction of 6 dBs.

The curve in Figure 1.7 illustrates the level to be added to the loudest of two sounds as a function of the SPL difference between the two sounds. In our case the difference is 0 dB; using the curve we see that we should add 3 dB to the loudest sound so the two sounds together will result in a combined SPL of 48 dB. A 3 dB increase represents a doubling of intensity. And a 6 dB increase is the doubling repeated for a quadrupling of intensity.

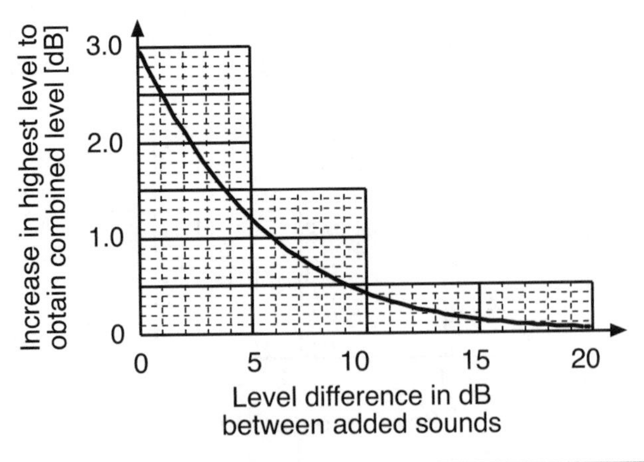

Figure 1.7 Assume that we have two sounds *A* and *B* having SPLs L_A and L_B respectively, and that $L_A > L_B$. The combined SPL of the two sounds is obtained from the curve that shows the number of dBs that should be added to L_A as a function of $L = L_A - L_B$.

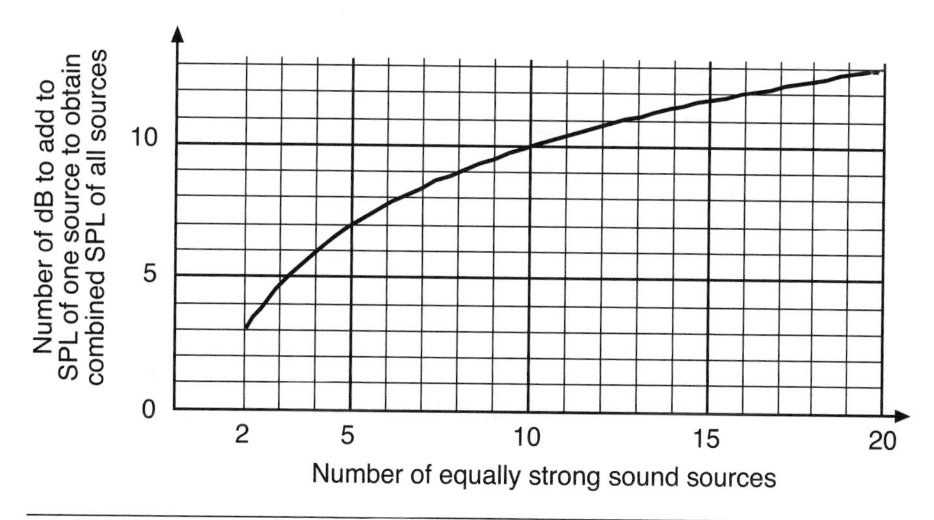

Figure 1.8 For the combined SPL of a number of equally strong sources, add the dB number given in the graph.

At times, one may need to estimate the combined SPL of a number of equally strong sources. Using the curve in Figure 1.8, it is easy to estimate the combined level of a number of sources.

1.5 SPECTRA OF VOICE AND MUSIC

The frequency contents of a sound is called its spectrum—in analogy with light. The spectra of sounds vary significantly from the simple—that can be generated by technical equipment—to the complex—voice and music.

Human voice is generated in two ways—voiced and unvoiced sounds. Voiced sounds—vowels—are generated by the vibrations of the vocal cords that act on the air stream pressed out from the lungs. Unvoiced sounds—consonants—are due to air blown out past the teeth or lips at a speed resulting in turbulence-generated sound such as consonants *f*, *s*, and *sh*. If the air is abruptly stopped, other consonant sounds such as *m* and *n* are generated; if abruptly stopped and released, sounds such as *p* and *t* are created.

By changing the shape of the oral volumes, for example by moving the tongue, one can change the properties of the acoustical filter and thus the spectra of vocal sound.

The voice has a wide frequency range, having frequency components from approximately 100 Hz up to approximately 10 kHz. Spectra are often shown as octave band spectra, and each data point represents the contribution to the level

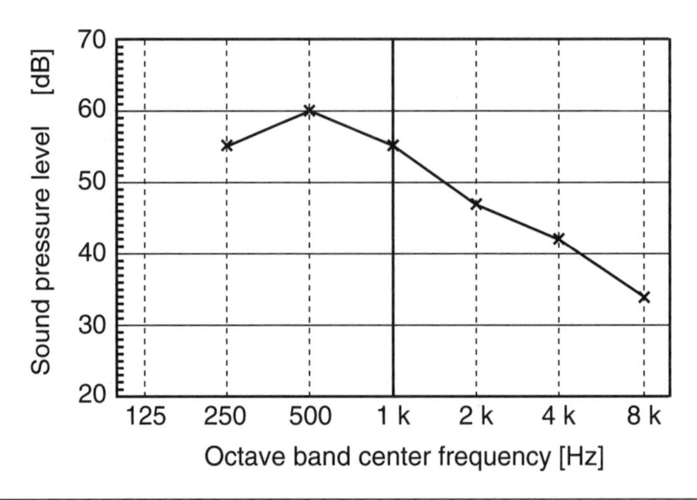

Figure 1.9 Analysis of speech shows that speech carries most power at relatively low frequencies as shown by these long term average octave band SPLs. The lines have been added to make a curve to improve legibility. Most of the energy is concentrated at relatively low frequencies due to the voiced sounds. The weak unvoiced sounds have predominantly high-frequency spectral content and are relatively important for speech intelligibility.

of all sound in an octave around the particular frequency. An example of the octave band spectrum of speech at a distance of 1 m in front of the talker, averaged for men and women, is shown in Figure 1.9. Its SPL is approximately 60 dB. A trained speaker can change the power output of the voice over approximately a 20 dB range. A whisper, which is also unvoiced speech, has an approximate sound power level of 30 dB.

Vocal sound energy is radiated primarily by the mouth and the nose openings; only the lowest frequencies are radiated by the chest. The sound shadow caused by the head will cause some directionality, as illustrated by the curves in Figure 1.10. The directivity of the voice is quite substantial at frequencies over 1 kHz; at this frequency there is more than 5 dB SPL difference between front and back radiation. Since most of the information in speech is at high frequencies, one should strive to speak in the direction of the listener.

Musical instruments may function in the same way as human voices; this is the case, for example, in reed instruments such as reed organ pipes and oboes. Other instruments use mechanical friction to set up vibrations in the instruments' surfaces that then radiate sound, examples are the string instruments such as violins and cellos. Music covers much wider frequency and SPL ranges than voice as indicated in Figure 1.11.

Musical instruments such as the pipe organ cover virtually the entire audible range of frequencies. The frequencies generated in playing the pipe organ are

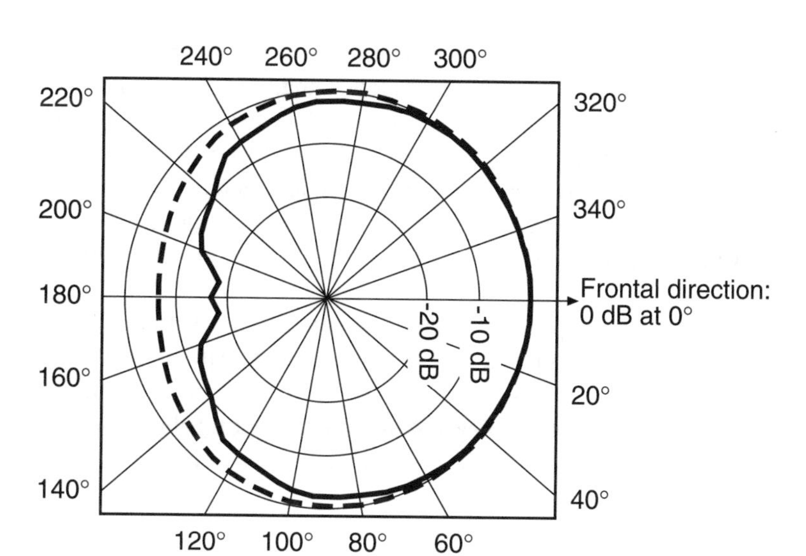

Figure 1.10 The curves show the relative octave band SPL of speech in various horizontal directions around a person. The solid curve represents the 4 kHz octave band whereas the dashed curve represents the 250 Hz octave band. The values must be combined with those given by the data in Figure 1.8 to give the actual values.

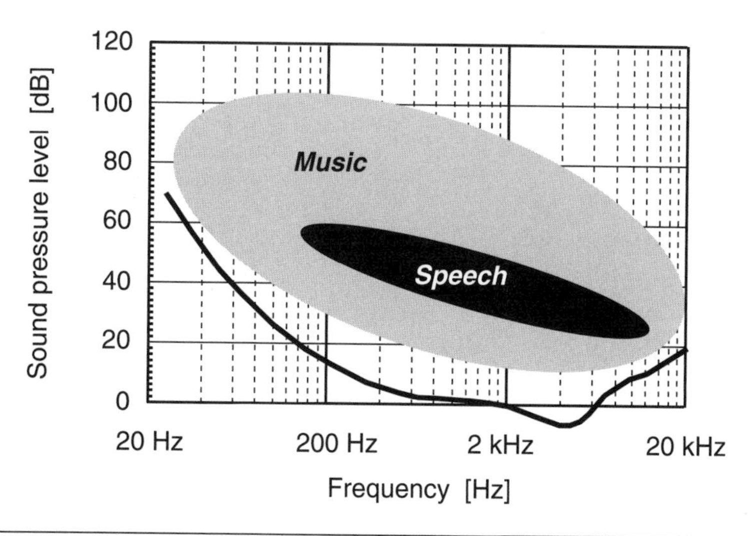

Figure 1.11 This graph shows the frequencies and SPLs of speech and orchestral music. The solid line shows the average threshold of hearing for young persons without hearing impairment.

the result of the air flow in the pipes. The air flow will energize resonances of the flue pipes (Diapason, Principle, so-called-string, and flute registers) and the reed registers (Trumpet, Krummhorn, Bombard, Vox Humana) of the organ. The frequencies of the sound being generated will depend on the size of the pipes; longer pipes are associated with lower frequency resonances. Because of the way sound is generated in the pipes, not only the fundamental resonance frequency of each pipe will be generated but also overtones and harmonics of the fundamental frequency.

For any note, the complex tone that is generated by the sum of all these components has a startup sound that is characteristic and that needs to be retained for the best possible sound. At the end of the note, the sound dies out at a pace characteristic of the room geometry and volume, the sound absorption due to its furnishing, the persons in the room, and other sound absorbers.

2

HEARING

2.1 BASIC PROPERTIES

Human hearing is remarkably sensitive and has a wide range. The signal processing that is active in hearing gives us the ability to detect extremely weak signals and signal patterns, even in noisy environments. For sounds to be audible, they must be strong enough to overcome the threshold of hearing. The threshold is determined by the mechanical construction of the ear and the sensitivity of the cochlea.

The frequency range in which we can hear sound is determined by the cochlea and has an approximate range from 20 Hz to 20 kHz for a person with normal hearing. There are significant variations in the upper limit that will depend on age, previous noise exposure, and other variants. Sounds having frequencies close to the lower limit will also be experienced as chest vibrations.

The strength of a sound is typically characterized by its sound pressure level (SPL). The SPL of sounds at the threshold of hearing is frequency dependent and approximately 0 dB at a frequency of 1 kHz. At this frequency, a doubling of the subjectively perceived loudness of sounds closely corresponds to a change of approximately 10 dB; for SPLs greater than approximately 40 dBs. The barely noticeable difference in SPL at this frequency is approximately 1 dB, which contributes to the popularity of dB level as a natural metric. The sensitivity of hearing is greatest in the range of 2 to 5 kHz and diminishes toward high and low frequencies.

Because of the varying sensitivity across the audible frequency range, shown in Figure 2.1, sound pressure is usually measured taking the variation into account in an approximate manner. Besides frequency, the loudness of a sound will depend

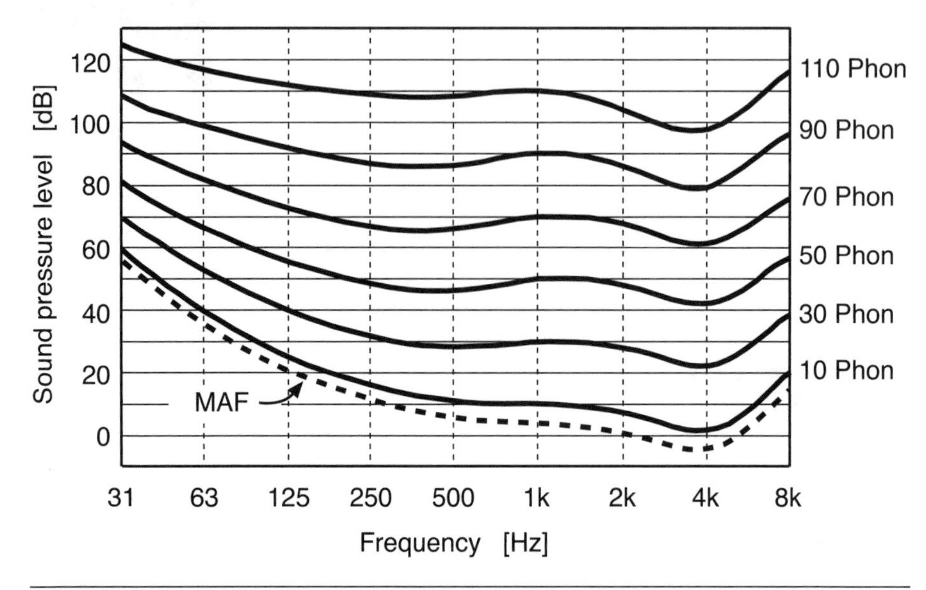

Figure 2.1 Some equal-loudness-contours for pure sinusoidal tones in the frequency range 31 Hz to 8 kHz. The curve marked MAF refers to the threshold of hearing for a plane wave of sound incident from the front. The numbers on the right of the equal-loudness-contours refer to the *loudness level* given in units of Phon. (Phon value is equal to the SPL value at 1 kHz.)

on the angle of incidence and the duration of the sound, properties that are not included in conventional acoustic noise measurements. Typically, omnidirectional microphones are used because their sensitivity is independent of the angle of incidence of the sound.

One usually characterizes the *strength* of a sound by measuring the SPL with a frequency sensitivity correction. Such corrections are achieved by so-called weighting filters. The weighted SPL values will typically correlate to the subjective determined loudness level values (in units of Phon) for sounds that have common wideband characteristics such as HVAC noise, voice, and music. The usual weighting filters are the *A*, *B*, and *C* filters. These give the sound level reading expressed in dBA, dBB or dBC, depending on the filter used. The frequency response curves of these filters are illustrated in Figure 2.2. Unweighted SPL values are frequently expressed in dBlin (linear dBs) to indicate that no frequency weighting was used.

The filter characteristics of the *A*, *B*, and *C* filters were chosen to resemble the shape of the equal-loudness-contours in the intervals 20 to 55 dB, 55 to 85 dB, and 85 to 140 dB, respectively. Because of the availability of data and considerable experience, the sound level expressed in dBA has become the most commonly used, particularly for the determination of the risk of noise-induced hearing loss.

Audible sounds at SPLs above approximately 130 dB are painful, but even sounds at much lower levels may be irritating. The *useful* range of hearing for

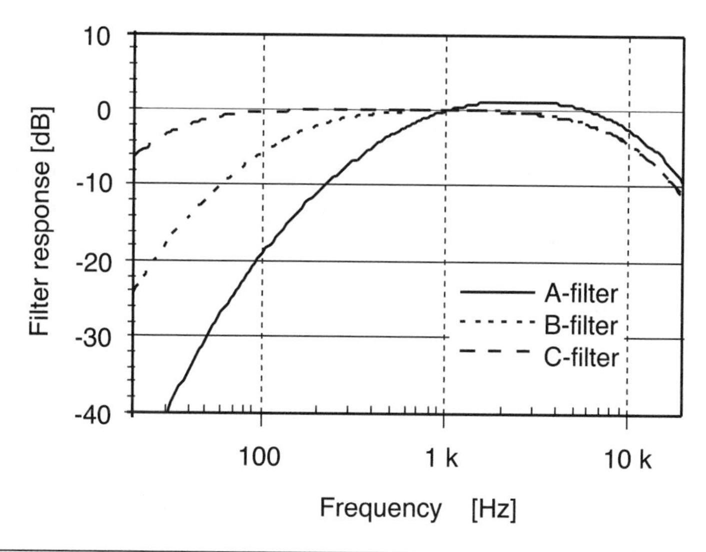

Figure 2.2 Frequency response characteristics for *A*, *B*, and *C* filters. (Negative values = filter attenuates signal.)

listening to speech—from the viewpoints of speech intelligibility and comfort—is 50 to 60 dBA.

This range of values assumes that the background noise levels of the listening environment are much lower. Comfortable listening requires background noise sound levels below 35 dBA, although, from the viewpoint of speech intelligibility, a level difference of 20 dB between speech and background noise sound levels is acceptable.

2.2 DIRECTIONAL PROPERTIES

The pinna (exterior ear) contributes much of the directional properties of hearing at high frequencies, but it is the head and the torso that are the primary influence on the directional characteristics of hearing. A trained listener, using only one ear, can sense the approximate direction from which a signal is coming. This is due to the changes in the spectral content of the sound at the ear, resulting from the influence of the head shadow affecting sounds at frequencies above 1 kHz.

One of the most important properties of our hearing system is that we have binaural hearing; we hear using two ears. This leads to a clarity and subjective perception of auditory space that cannot be accomplished when listening with the use of only one ear. This is due to the phenomena of localization, noise suppression, and decreased masking.

The better localization ability is the result of a binaural hearing system that compares the arrival times and SPL differences between the signals from the two ears. Since hearing also analyzes the patterns and information contents of what we hear, it is capable of tracking characteristic sounds and their origin, even when they appear to come from many directions such as in echoes and reverberation. Provided that these are not too strong relative to the direct sound from the source, it is the arrival direction of the first instance of a sound that determines from where it appears to be coming. This is called the *precedence effect* (*cocktail party effect*), which is an important property of hearing when listening to sounds in reverberant spaces.

2.3 MASKING AND CRITICAL BANDS

A sound, the masker, blocks the perception of another sound, the maskee—this is known as masking. Masking can take place both in the time and frequency domains. Simplified, one can regard masking as the shift of hearing threshold in the presence of another sound such as a tone or noise. The masking effect is greatest when the masker and the maskee are close to one another in frequency. The effect is primarily active upward in frequency. The masking level is the level by which one sound needs to be increased in order to be heard in the presence of a masker. From the viewpoint of speech intelligibility, low-frequency maskers are the most disturbing because they will mask much of the most important frequency range of speech. The masking effect in the time domain is discussed in Reference 2.5.

2.4 HEARING IMPAIRMENT

Sounds begin to become painful at SPLs of about 130 dB. Long-term exposure to sounds at levels of 85 dBA or more, however, is considered in many countries to be the limit at which serious hearing impairment is possible and hearing impairment is irrevocable.

Impaired hearing particularly influences our ability to understand speech and to enjoy music. Both age and noise exposure such as prolonged listening to loud music, contributes to hearing loss. Musicians often suffer hearing impairment. Figure 2.3 shows some data for the hearing loss in the population as a function of age (see Reference 2.2).

Most elderly people suffer from age-related hearing loss (ARHL) or presbycusis that usually takes the form of high-frequency hearing loss and is associated with a loss of hair cells that are sensitive at high frequencies. It is a result of general wear-and-tear due to noise exposure. Health-related conditions, such as high cholesterol levels, and poor health habits that influence the cardiovascular system,

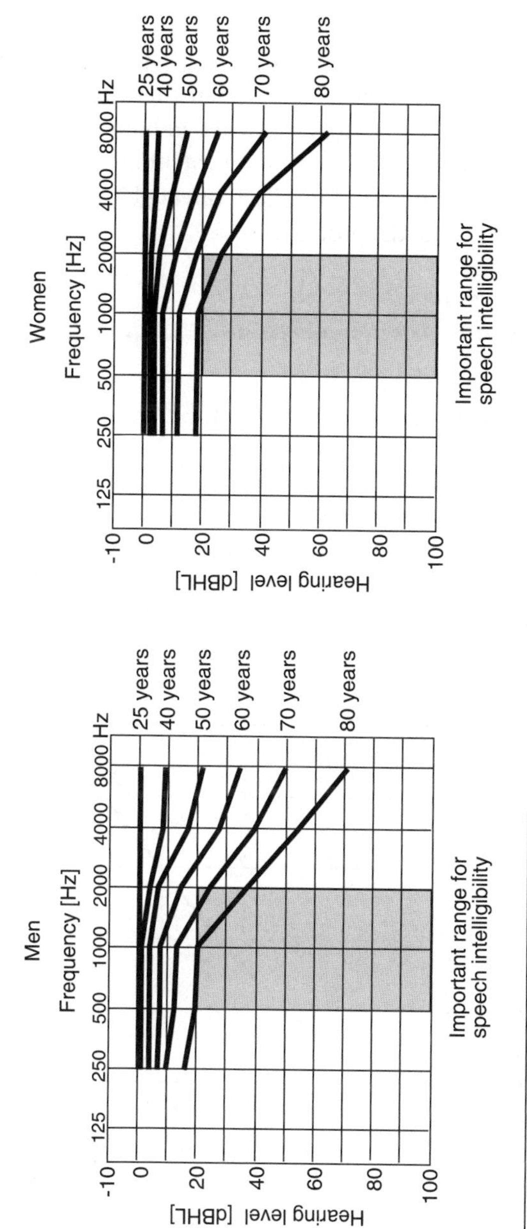

Figure 2.3 Hearing threshold shift due to ARHL for various age groups (for men and women respectively).

such as smoking, also seem to influence ARHL. Curves characteristic for hearing loss due to presbycusis are shown in Figure 2.3.

Typically, noise-induced hearing impairment or occupational hearing loss, first affects the frequency range around 4 kHz. Thus, it is the high-frequency consonants that suffer the most, whereas the fundamentals and harmonics of vocal sounds are less affected by hearing loss. However, in most Western languages, the consonants are the major carriers of speech information and the frequency range of 1 to 4 kHz is the most important for speech intelligibility.

These examples illustrate the problem:

> *Text with voiced sounds only:*
> ••e •o•• o• •ea•i•• •o•• •o••e••a•io• i• •i••.
>
> *Text with unvoiced sounds only:*
> Th• c•st •f h••r•ng l•ss c•mp•ns•t••n •s h•gh.

As illustrated, the information contained in the sentence is best retained by the consonant sounds. Good acoustics for consonant sounds, that is, sounds that have frequencies in the range 1 to 4 kHz, are thus crucial for good speech intelligibility. Figure 2.4 depicts how different frequencies in speech contribute to intelligibility (see Reference 2.1). Note that good speech intelligibility is not synonymous with

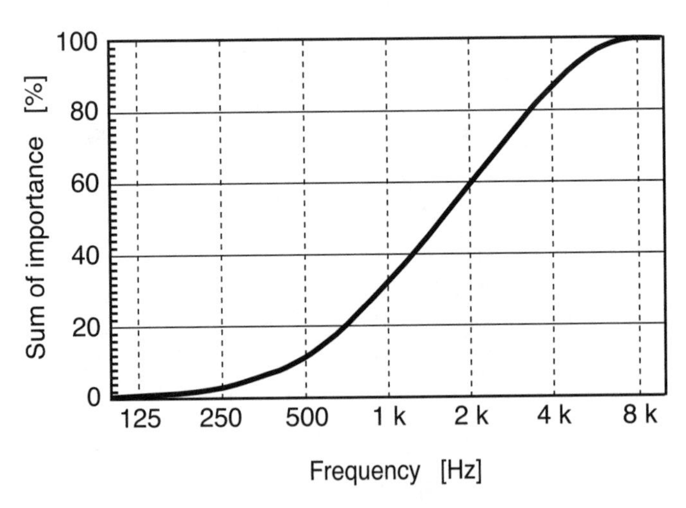

Figure 2.4 Most of the energy of speech is concentrated at relatively low frequencies due to the voiced sounds. However, the unvoiced sounds have predominantly high-frequency spectral content and are relatively more important for speech intelligibility. The curve shows the importance of various frequencies to speech intelligibility. Note that the frequently range between 1 and 4 kHz is the most important for intelligibility (the curve has the highest slope).

good speech (sound) quality. Speech quality is the way the speech *sounds* and is influenced by various distortions, including some from electronic speech reinforcement and/or communications equipment.

Note also that because the information content in speech is dominant at high frequencies (see Figure 2.4), ARHL (see Figure 2.3) is serious for speech intelligibility. A sensitivity drop of 20 dB roughly corresponds to a tenfold increase in distance from the sound source. In a reverberant environment, such as a typical worship space, this may make speech quite unintelligible at a distance.

A peculiarity of occupational hearing loss is that it frequently affects one ear more than the other. Violinists, for example, usually have damaged hearing in their left ears. If one ear is damaged, the binaural unmasking generally available to us will not function effectively, and it will be more difficult to extract the direct sound signals from the reverberation noise.

2.5 EFFECTS OF MASKING IN TIME

The reflected sound from walls, ceilings, and floors, not to mention humans and objects, competes with the direct sound for the attention of the listener. Our hearing separates sounds in time and in space. Because the spatial resolution is poor due to the large wavelengths of many of the sounds that we hear, time resolution is also important. If sounds follow one another at a rapid pace in a reverberant room, as indicated in Figure 2.5, we will not hear the reverberant sound separately, and the reflected sound may reinforce the direct sound or interfere with subsequent direct sound, depending on both how long the reflected sound energy persists in the room and the relative levels of the first and the later sound.

Speech sounds are made up of wave packets; there are typically 7 packets per second in normal speech so that a word consisting of two consonants and a middle vowel may last nearly half a second. Hearing analyzes sounds for patterns. Familiar sounds, including speech sounds, are immediately recognized. Hearing is also capable of combining this pattern analysis with a spatial analysis using the precedence effect discussed previously.

Hearing separates the reflected sounds into two groups; those that arrive shortly after the direct sound from the sound source and those that arrive with a delay. The limit corresponds to approximately a path length difference between the direct sound and sounds at the limit of 17 to 20 m (60 to 70 ft), that is, 50 to 60 milliseconds. Remember, the speed of sound in air under normal conditions is 344 m/s (approximately 1129 ft/s).

Sounds that arrive shortly after the direct sound affect the color and spaciousness of the direct sound and creates an apparent auditory source width of the sound source. Typically, to our ears, the source will seem to be much greater than

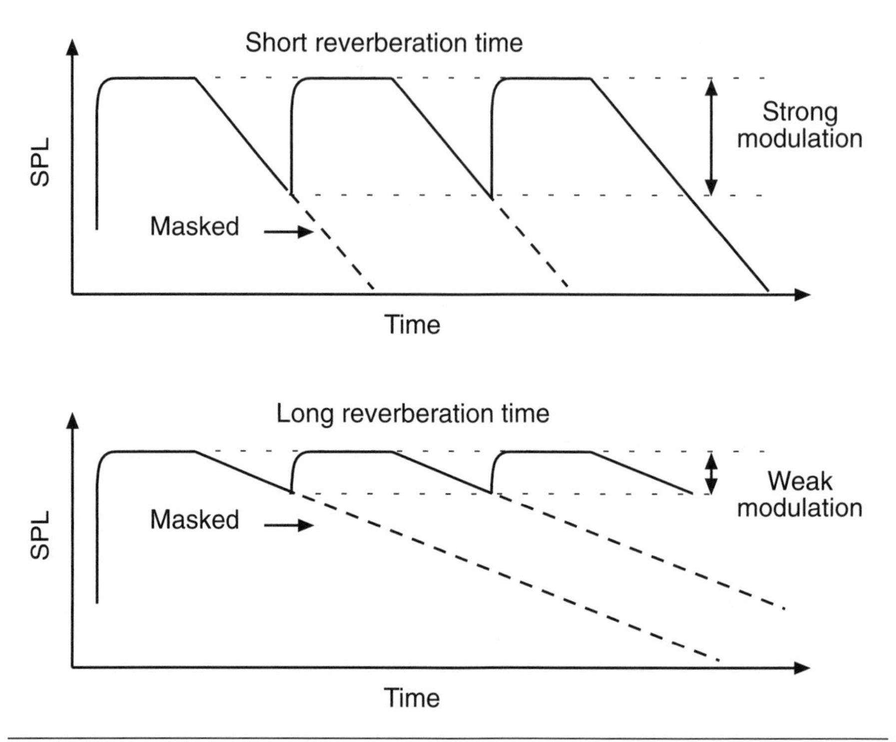

Figure 2.5 Reverberation by previous sound masks later sound. The modulation depth decreases, as a result, and less detail is heard.

its viewed size. This is particularly noticeable with the voice of a single singer, for example. It is the sound arriving at our ears from the side walls (often called lateral sound by acousticians) that is the major contribution to the feeling that the sound has an aural spacious quality. In a room in which one wants to achieve *acoustical intimacy*, the earlier sound needs to be fairly strong.

Sounds that are delayed due to their traveled path lengths, and thus arrive after the mentioned time limit, generate the sensation of reverberation. Reverberation competes for our attention with the direct and early-reflected sound. From this viewpoint, the reverberation can be considered a form of noise—unwanted sound. Some reverberation is appreciated because it gives *fullness* to speech, but for greater speech intelligibility reverberation time (RT) should be kept to less than 0.8 to 1.0 s or the ratio of all early to all reverberant sound should be at 10 dB. This means that one word following the next will be attenuated by some 20 dB and the resulting signal-to-reverberation ratio will be sufficiently high.

Music typically has different temporal composition than speech. Many musical instruments have sounds of longer duration than speech sounds. The onset part of a typical flue organ pipe sound needs nearly 100 ms to develop, approximately the same time as a speech sound, but it will last much longer due to the slow tempo of the music.

For music (including singing to some extent), much longer reverberation times are needed than for speech to give fullness and *body* to the music. Reverberation is essential to the formation of emotion and the feeling of spaciousness also. Some Christian clergy use the phrase *a sense of mystery* in describing the desirability of reverberation in their worship space.

Even though traditional Muslims and Jews do not regard reverberation as essential to their respective religious worship, they still consider reverberation an essential part of their music or chant appreciation and view it as a natural part of their worship environment.

3

ROOM ACOUSTICS FUNDAMENTALS

3.1 PROPAGATION AND THE REFLECTION OF SOUND

Sound waves will spread out from the source. If the waves do not meet any acoustical obstacles, the intensity will drop as the wave covers a larger area. For small sound sources, the pressure will diminish inversely to its distance away from the source.

Sound is reflected when it reaches surfaces that have acoustical characteristics different from air. Some of the sound that falls onto an object will be absorbed, some will be transmitted, and some will be reflected as shown in Figure 3.1.

When sound is reflected by walls, floors, windows, or other objects, most of the sound will be reflected specularly as long as the surfaces are large compared to the wavelength of the oscillation and as long as they have low sound absorption. Since practical building constructions have limited mass and stiffness, they will be put into vibration by the incident sound. Such vibration will result in sound being radiated by other surfaces in the building that are reached by the vibration. This results in a lack of sound insulation, but seldom leads to high sound absorption except in the case of windows.

If an object, for example a light fixture or a pillar, is in the way of a sound wave, and the object is small compared to the wavelength of sound, most of the sound moves past the object as indicated in Figure 3.2.

Experience tells us that unless a mirror is flat and clear, the mirrored image will be distorted. Similarly, if a surface is uneven, the sound will be scattered. One

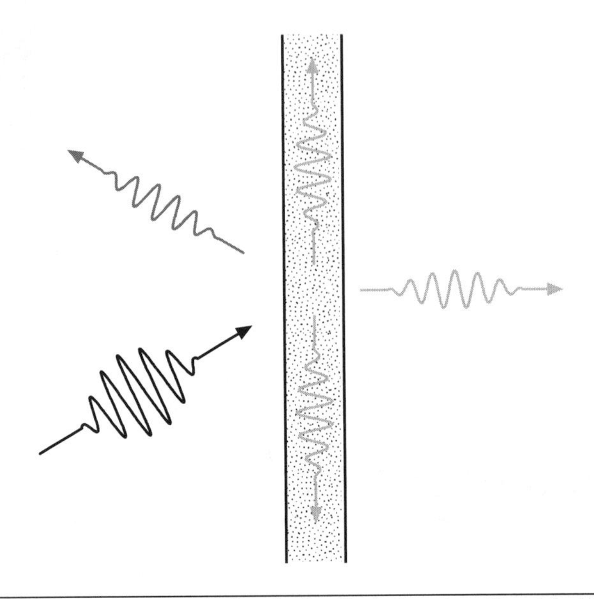

Figure 3.1 Sound can be transmitted, absorbed, reflected, and scattered (not indicated) by a wall.

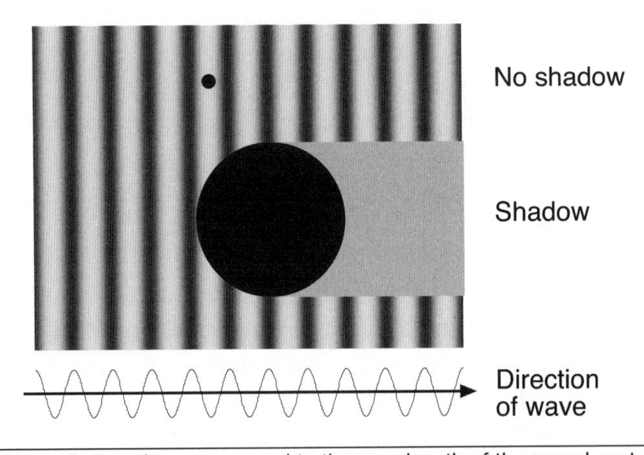

Figure 3.2 Objects that are large compared to the wavelength of the sound cast shadows.

can think of the sound leaving the source as rays that carry the energy of the sound. As the rays move farther and farther away from the source, they will be at a greater distance from one another and thus more sparse. This is equivalent to the reduction of sound amplitude discussed earlier.

3.2 OUTDOOR SOUND

Let us consider a person speaking outdoors. We can consider the speech as small wave packets consisting of syllables. These are sent out as a multitude of rays as illustrated in Figure 3.3. We assume each ray carries part of the speech sound sent out by the source and the strength of each ray being inversely proportional to the number of rays. The SPL associated with the wave packet drops 6 dB per distance doubling as discussed previously.

The wave amplitude also drops as a function of sound frequencies and air humidity. The peak of the damping effect due to the humidity in the air will be found to occur at lower levels of relative humidity. This damping increases rapidly with frequency and is important to consider for reverberant sound, because

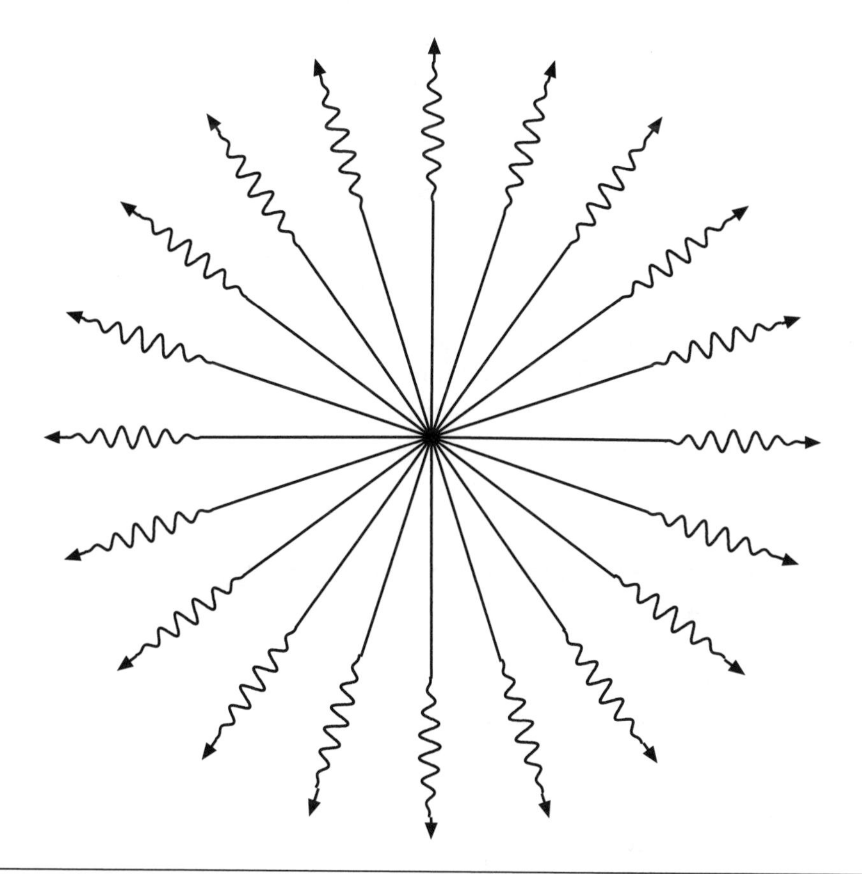

Figure 3.3 The voice can be considered a source of a multitude of rays. Each sound ray contains a wave packet representing part of the source output. The rays to the front of the head are stronger than those to the rear because of the shadow cast by the head.

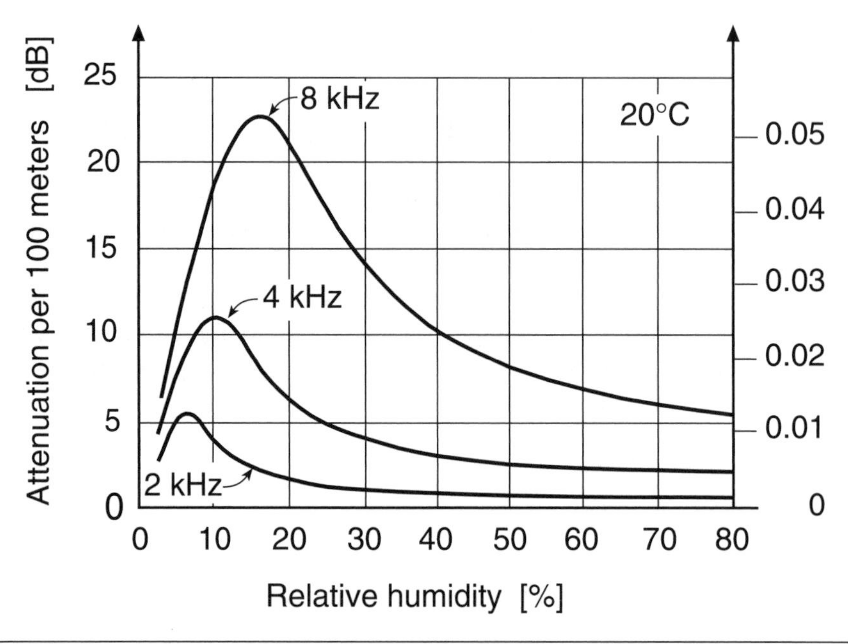

Figure 3.4 Damping by air humidity is frequency-dependent. This graph shows the attenuation of a plane wave of sound at 2, 4, and 8 kHz in dB after it has traveled 100 meters. The attenuation coefficient *m* can be calculated by dividing the dB value by 434 (right scale, metric units assumed).

reverberant sound will travel long distances before it becomes inaudible. In large spaces such as cathedrals, the sound absorption due to the humidity of the air will dominate over all other sound absorption at high frequencies. The curve in Figure 3.4 shows the attenuation of sound for different values of the relative humidity for a frequency of 8 kHz.

3.3 GEOMETRICAL ACOUSTICS, WAVE PACKETS, AND SOUND RAYS

An intuitive method to analyze the initial response of a room to a sound is geometrical acoustics; we regard sound as traveling like rays. Geometrical acoustics is the basis for many software models for predicting the sound propagation in rooms as well as outdoors.

In geometrical room acoustics we use the same modeling principles as in optics, that is, using the ray tracing method and the method of mirror images to study the propagation of sound. The method is frequently extended by further acoustically relevant approximations. Geometrical room acoustics is useful when studying the behavior of sound both indoors and outdoors as long as its shortcomings are known and it is used appropriately. In this method, surfaces much larger than the wavelength of sound reflect. Surfaces and objects the same size or smaller than the wavelength scatter (diffuse) sound. The energy in a ray is split up into a multitude of new, lower intensity rays, as indicated in Figure 3.1.

The path of each ray is a straight line unless the ray hits an object. During its travel, a ray may be reflected as well as bent, depending on the presence of sound-reflecting objects or wind and temperature changes. Acoustic mirrors reflect sound at the same angle as it is incident, as in optics. A mirror image is formed by a perfect mirror, and we can consider a reflected ray as emanating from such a mirrored source (often called a mirror image).

In applying ray tracing, we can use one or both of two methods: 1) send out rays and follow each ray along its path or 2) find the mirror images of the source and then draw rays from the mirror images of the source to the place of interest.

The case of two opposing mirrors in Figure 3.5 is simple to follow. Note that the number of times a reflection has been mirrored, or a ray path reflected, is called the mirror image or ray-reflection order.

Because it becomes cumbersome to keep track of all the mirror images, the more direct Method 1 is usually applied. We detect the arrival of rays at a seat in the auditorium by using a test sphere so that we can predict the temporal and spatial patterns of rays at that listening position.

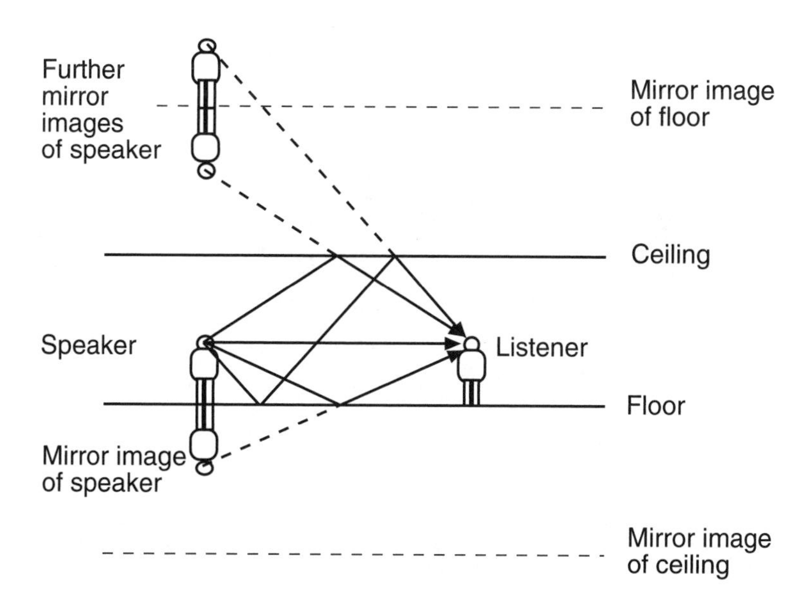

Figure 3.5 Rays can be thought of as coming from both the real source and the mirror images of that source.

When surfaces are curved or rough the ray tracing method excels. A curved surface may be subdivided into smaller surface patches and analyzed, assuming that the surface is large as compared with the wavelength. For rough surfaces we subdivide the energy of an incident ray and assume it is distributed in many directions as new rays, each carrying a fraction of the energy of the original ray.

Clearly geometrical acoustics is primarily useful for the study of the behavior of the early sound distribution in rooms. In practical rooms, the unevenness of the walls, the inaccuracies of wall positions, and the sheer multitude of rays or images make it difficult to analyze the conditions by hand after the first or second order of reflection.

Figure 3.6 presents an example of the use of a simple mirror system and a laser to trace rays on a chapel plan. To trace more rays, it would be necessary to use computer modeling. Higher reflection orders are easily handled by using dedicated computer software. Modeling using the software for ray tracing typically assumes that a large number of rays (1000 to 1,000,000) are being sent out from the sound source. The sound source can be assumed to be omnidirectional or directional as in the cases of the voice, many musical instruments, and loudspeakers. An example of ray tracing results from a church model is illustrated in Figure 3.7.

In practice, sound reflection time patterns of rooms are not as simple as one might be led to believe by simple geometrical optics. As shown in Figure 3.8, the measured impulse response of a real room is complicated and cannot be calculated exactly. This is due to the simultaneous presence of many physical phenomena such as scattering, diffraction, complex surface impedance, and source

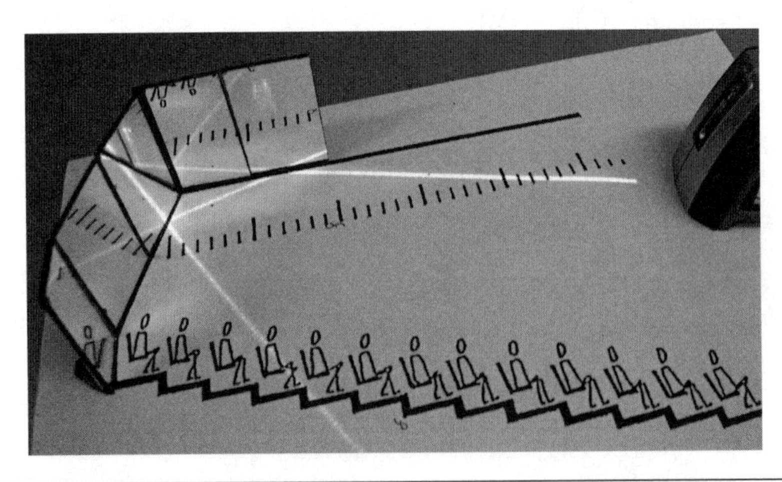

Figure 3.6 Simple ray tracing can be done with mirrors on a room drawing using a laser tracker.

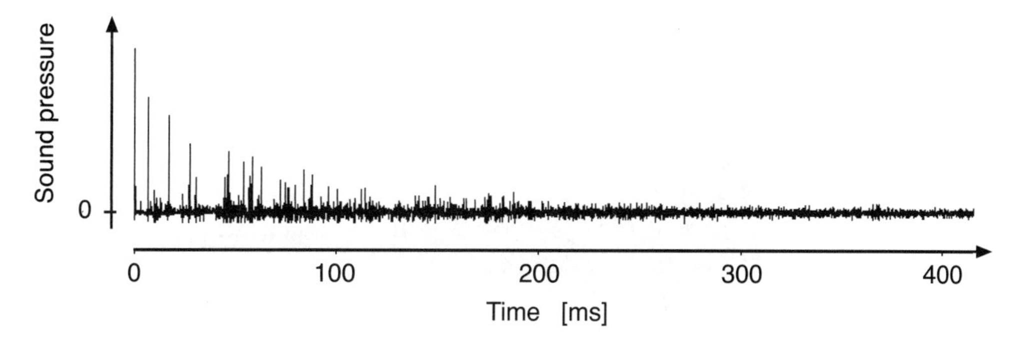

Figure 3.7 An example of the calculated sound reflection time pattern between the priest's pulpit and a listening point farther back in the church shown in Figure 6.1 (sound pressure vs. time).

and receiver responses. The energy behavior, however, can be quite successfully simulated.

For any venue, the time gaps between the various rays arriving at the listener will, of course, vary with listener and sound source location. The time gap between the direct sound and the major first sound reflection is defined as the initial time gap (T_I) and influences the subjective feeling of intimacy. The early-reflected rays should be wideband, that is, have full frequency ranges. One should avoid strong overhead reflected rays appearing in the same vertical plane as the direct ray.

Because hearing cannot differentiate between signals having the same interaural delay time, excessively strong overhead reflected sound will seriously reduce the sense of spaciousness. However, in applying remedial measures to badly designed spaces, this has often been a quality sacrificed in the quest for clarity and, reasonably, even early sound distribution through the seating area. New designs can avoid this problem.

The unpleasant lack of spaciousness can be avoided by insuring the overhead reflecting surfaces are diffuse and/or insuring sufficient lateral-reflected energy within the first 30 milliseconds coming from the sides of the room or other overhead reflectors. A design using translucent reflectors is shown in Figure 7.26 (top center of photo).

3.4 REFLECTION OF SOUND

There are few surfaces in buildings that can be considered perfectly reflecting at all frequencies of hearing in the sense of geometrical acoustics. At low frequencies, the room dimensions may simply be too small for the room surfaces to be

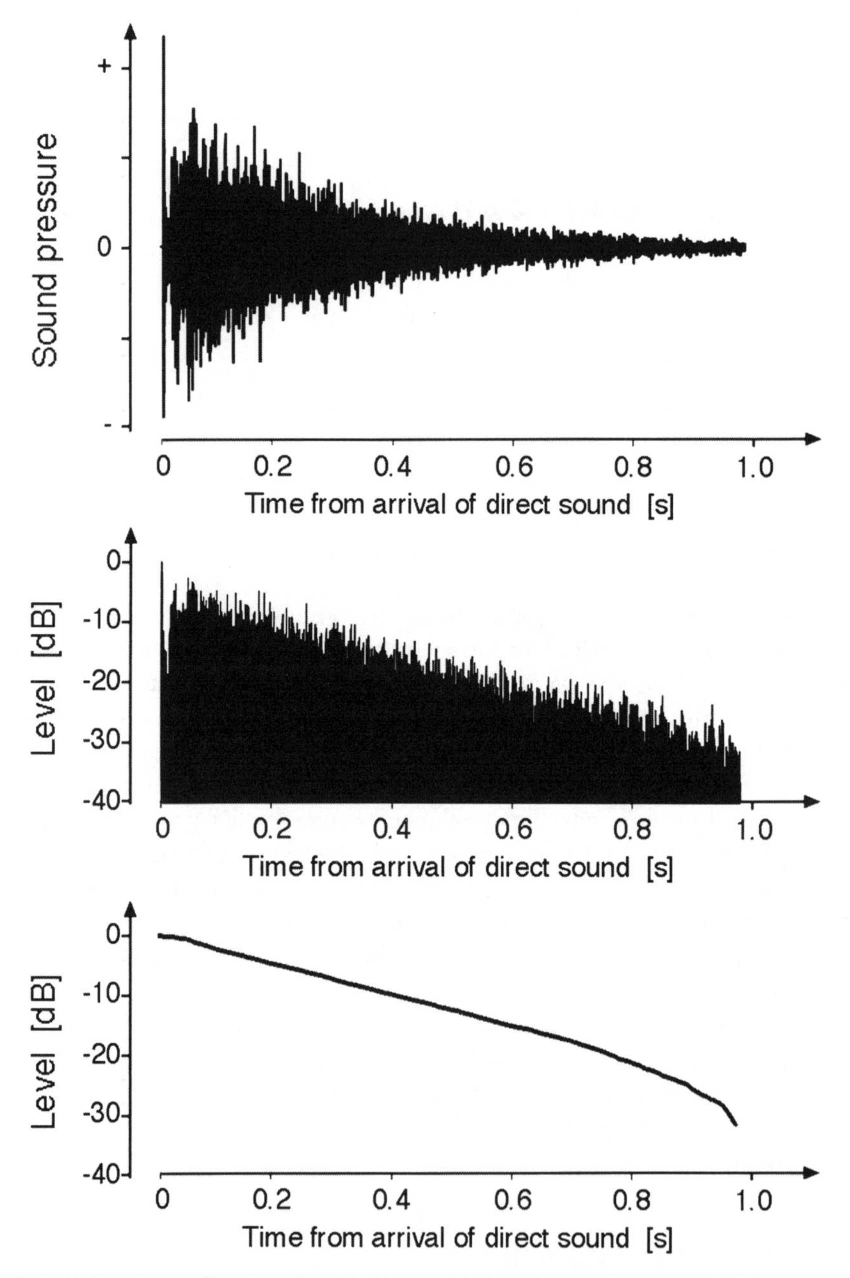

Figure 3.8 Three ways of representing the same measured sound decay characteristics in the church modeled in Figure 6.1. The top curve shows the sound pressure drop (impulse response), the middle curve shows the SPL drop, and the bottom curve shows the reverberation curve obtained by reverse summation of energy in the top curve.

of sufficient size compared with the wavelength; whereas at high frequencies, the irregularities caused by poor workmanship and unavoidable lack of precision are such that the surfaces are more opaque than mirroring.

When sound from a speaker hits the floor, it will be reflected. Some of the energy is lost to the floor because vibrations are set up in the floor—unless its construction is heavy—and some is converted to heat. A fraction of the energy, symbolized by the Greek letter α, is lost. This is called sound absorption and the fraction is called the sound-absorption coefficient and will be a number between 0 and 1. The fraction will be dependent on the acoustic properties of the surface material—its size and geometry. The absorption will vary with the frequency of the sound; some surfaces and building constructions have sound absorption that is fairly constant over the frequency range. Others will have sound absorption that varies considerably with frequency such as draperies (efficient absorbers at high frequencies) and lightweight wooden constructions (efficient absorbers at low frequencies).

The sound will be reflected at the same angle that it is incident on the surface, as long as the surface is plane and much larger than the wavelength of sound. If the surface is small, the wave packet's energy will be diffused in many directions as discussed previously.

Because of the way the absorption coefficient α is defined, the effective sound-absorbing area A is given by $A = \alpha S$, where S is the area of the sound-absorbing surface. A list of typical sound-absorption coefficients is shown in Table 4.1.

Not only the surfaces of the room, for example floors, walls, ceilings, absorb sound but also objects in the room, including people and furniture. As a rule one can say that anything that has a porous surface will absorb some sound. The sound absorption by people will be dependent on dress and will typically vary to some extent between summer and winter.

The diffusion coefficient, δ, is the fraction of the reflected sound power that is being reflected diffusely. The diffusion coefficient will be a number between 0 and 1. Only the sound that is not being absorbed can be reflected either specularly or diffusely (see Figure 3.9). Construction, craftsmanship, or material properties are a few reasons why there will always be some diffusion of sound by a surface.

The assignment of a diffusion coefficient to a surface is currently based on experience and informed guesswork. The smoother a surface is, the smaller the diffusion coefficient. The smoothness is related to the ratio between the effective size of surface irregularity and the wavelength of the sound. The diffusion coefficient will generally vary considerably with frequency and building construction. An example of the typical assumed behavior of the diffusion coefficient is given in Figure 3.10.

An important effect of scattering is that sound will be reflected more often per time unit. Thus, the effectiveness of sound absorption by the surfaces of the room will increase typically resulting in shorter reverberation. Scattering will

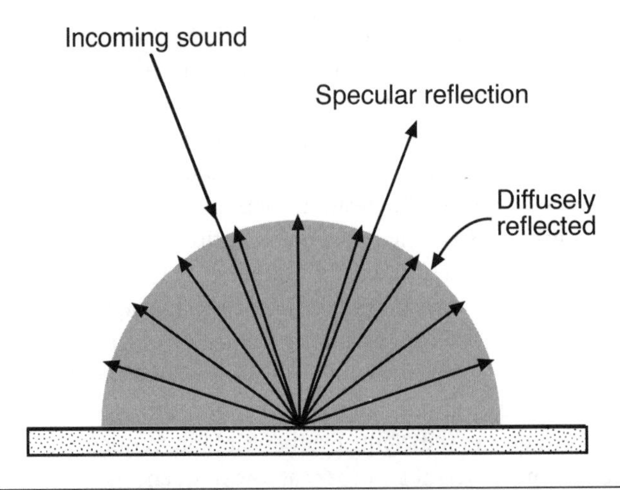

Figure 3.9 Sound is typically absorbed, reflected, and scattered (diffused) at a surface. The part of sound that is reflected (1-α) is either reflected specularly (1-δ) or diffused (δ).

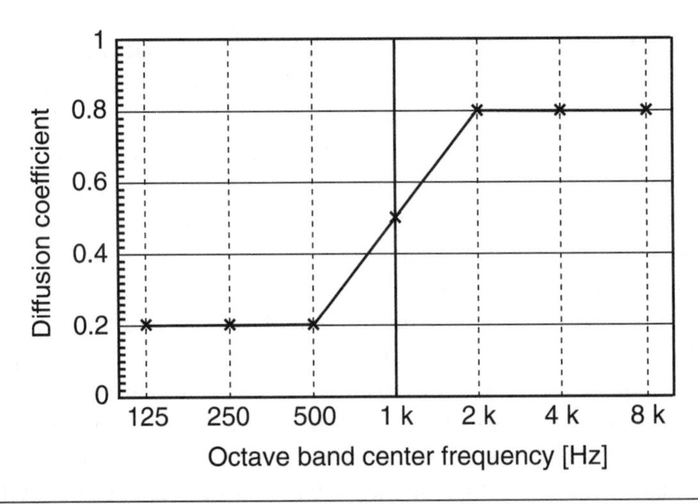

Figure 3.10 Typical assumed behavior of the diffusion coefficient.

also remove *hardness* from the sound of a room, in a similar manner as a sound absorber but without the *lifelessness* associated with an absorber.

3.5 SOUND DECAY IN ROOMS AND SABINE'S EQUATION

Because of the significant range of sensitivity of human hearing and the low speed of sound, we can hear how sounds decay in a room. It will typically last a second or two until a sound has died away and has become inaudible. The louder the sound, the longer it will remain audible.

The time that it takes for SPL to drop by 60 dB is taken as a measure of the damping characteristics of the room and is called the reverberation time (*RT*) of the room. The greater the volume (*V*) of the room in proportion to its surfaces, the more sound energy can be stored in the room's air. The approximate reverberation time is given by Sabine's equation shown in Equation 3.1 for imperial units:

$$RT = T_{60} \approx \frac{0.49V}{\sum_i A_i} = \frac{0.49V}{\sum_{i=1}^{N} \alpha_i S_i + \sum_{j=1}^{M} A_j + 4mV} \quad \text{(Imperial units)}$$

$\Sigma\alpha S$ is the total sound absorption in Sabins, where one Sabin is a square foot of total absorption, $\alpha S = 1.0$. In the more complete formula, V is room volume in cubic feet, α_i is the sound-absorption coefficient, and S_i is the area in square feet. For objects that cannot be assigned a surface area easily (individual persons or objects in the room) the absorption is A_i Sabins.

Sabine's equation is shown in Equation 3.2 for metric units:

$$RT = T_{60} \approx \frac{0.161V}{\sum_i A_i} = \frac{0.161V}{\sum_{i=1}^{N} \alpha_i S_i + \sum_{j=1}^{M} A_j + 4mV} \quad \text{(Metric units)}$$

$\Sigma\alpha S$ is the total sound absorption in metric Sabins, where one metric Sabin is a square meter of total absorption, $\alpha S = 1.0$. In the more complete formula, V is room volume in cubic meters, α_i is the sound-absorption coefficient (same as above), and S_i is the area in square meters.

The influence of air humidity is accounted for by the parameter m. The value of m is determined by the temperature and humidity of the air (and on the choice of metric or imperial units). For values of m at normal room temperature see Figure 3.4 or, for a wider range, see Reference 8.4.

Wallace Clement Sabine taught physics at Harvard and developed Equation 3.2 in analyzing the acoustic problems at a lecture hall there. He then employed this formula in advising architects McKim, Mead, and White on the design of Boston's Symphony Hall, This formula has proved successful in the design and correction of many concert, theater, and opera performing spaces. Additionally, it is effective in most lecture rooms and in worship areas. Furthermore, large railway, bus, and terminal concourses, hotel ballrooms and atriums, and similar spaces can be designed confidently using this formula. But wide-area, low ceiling spaces, such as many supermarkets, and long, narrow corridors, including subway stations, require specialized formulas outside the discussion of this book (see Reference 3.1) or the use of computer modeling as discussed in Chapter 6.

The formula is based on a statistical model for the sound propagation in the room. It assumes that the mean sound-absorption coefficient is less than 0.3, that the room does not have an extreme geometry, and that the sound-absorption properties of all room surfaces are approximately equal.

If one or more of these assumptions is not valid, one must estimate the reverberation time using a ray tracing model or, possibly, the room can fit the criteria for the use of one of the other formulas. Such models can deliver good correspondence between reality and modeling if handled by an experienced user.

The reverberation time is the most used metric for the relative suitability of a room's acoustics for a particular purpose. Because absorption coefficients are frequency dependent, the reverberation time will vary with frequency. Typically, audiences and performers prefer that reverberation times are fairly similar over the range of frequencies between 250 Hz to 4 kHz for any particular purpose.

An increase in reverberation times at low frequencies gives *warmth* to a room for music presentation but may result in voices sounding *boomy* if carried to the extreme. A reverberation time at the highest frequencies that is as long as that at mid frequencies (0.5 and 1 kHz) will give the room a brilliant and a too *bright* character. A room having a short reverberation time is usually referred to as acoustically *dry* and uninteresting. In spite of this, such reverberation times are usually desirable for small rehearsal rooms and studios. Its opposite is an acoustically *live* room.

3.6 REVERBERATION TIME AND REVERBERATION CURVES

The reverberation time can be measured using either analog or digital equipment. It is measured using the reverberation curve that shows the decay of SPL in a room after the sound sources have become quiet. Rooms that have a complicated geometry may have parts with different damping and reverberation times. This effect will make the reverberation curve have double or more slopes. An example of such a curve is shown in Figure 3.11.

The reverberation curve can be measured using music but, typically, specialized equipment is used. Common desired reverberation times are shown in Figure 7.13.

3.7 SPATIAL PROPERTIES OF REVERBERANT SOUND

One can easily form an opinion on the importance of the spatial qualities of reverberation by considering the difference in a musical experience when listening to a stereo sound system at home and attending a live performance in a concert hall. Most music recordings will have reverberation added by the sound engineer.

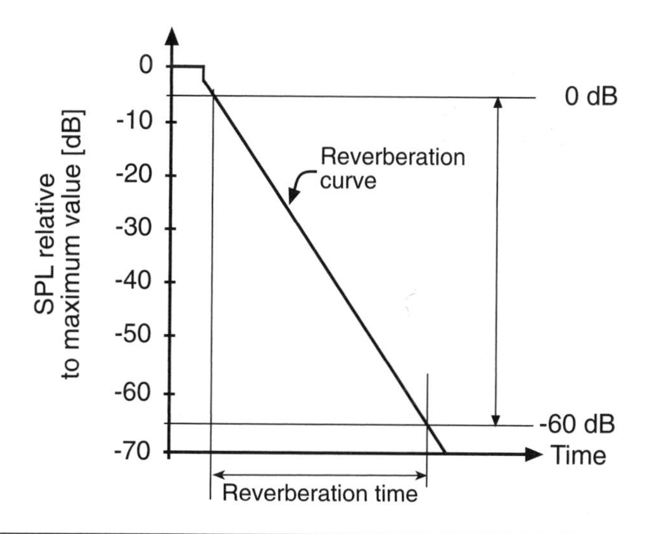

Figure 3.11 Determination of reverberation time. The reverberation time is defined as the time it takes for the SPL to drop by 60 dB after the sound source has stopped.

It can be natural, picked up by the microphone from the hall or studio in which the recording is made, or added using a reverberation unit in the studio. When listening to the sound from such a recording, at home or in the car, unless one has a sophisticated ambience and surround system with multiple loudspeakers and specialized signal processing, one immediately notices that the reverberation does not sound natural. It appears to come from either the loudspeakers of the stereo system or from some place in between the loudspeakers. However, in a large and reverberant room, the reverberation takes on a spatial quality that could be called *celestial*, not only because of the way in which it slowly decays, but also because it appears to come from all directions around the listener.

The spaciousness of the reverberation is due to the way that hearing processes the signals from the ears. Hearing does compare the reverberation from the two ears (due to a particular originating wave packet) and, if the comparison turns out poor, that is, if the sound on the left and right ears are significantly different (because of the different paths traveled), the feeling will be that of *spaciousness* and *envelopment*. We feel that we are immersed in a space filled by reverberant sound.

It is beneficial to strive for many reflection-incidence angles because this contributes to the feeling of being immersed in the reverberant sound. To obtain good subjective diffusitivity, direct overhead reflections should be somewhat weaker lateral reflections (but overhead reflectors can be designed to give lateral reflections). The side walls of a room should be somewhat scattering by using shelves, balconies, or unevenly placed sound-absorbing patches or objects, for example, having sizes in the range of 0.05 to 0.5 m (ca. 2 in to 2 ft).

When listening to music in a concert hall or worship space, having the feeling of being immersed in luscious reverberant sound is what gives the magical quality to the musical experience, particularly toward the end of a full chord when the reverberation is allowed to decay into complete silence and, especially when there is a moment of silence after the final chord before the applause. Thus, one cannot accept noise in a concert hall or worship space because it distracts from the experience of sanctity. One wants the sound level of the background noise, particularly of a large room, to be low. The more the *intelligible* character of a background noise, the more distracting the noise will be.

The spatial distribution of the early-reflected sound is important particularly for creating the impression of *auditory source width.* One should strive for sound reflection angles close to the horizontal plane to maximize the advantages of our binaural hearing. It is advantageous if the directional distribution of the early reflections be such that the sound field appears to be approximately, but not exactly, symmetrical when listening.

3.8 LOUDNESS OF SOUNDS IN ROOMS

Normally, beyond some point away from the source, the reflected sound energy of the room will dominate the sound. That is because beyond this point the direct sound will be weak compared to the reverberation that consists of all the sound bounced back from the surfaces of the room. The direct sound may be reinforced by early reflections. The larger the sound absorption in the room, the weaker the apparent sound source will be to our hearing.

When listening to common sound sources in nonreverberant environments, one is struck by how weak they sound. For adequate strength some reverberation is necessary but, at the same time, the reverberation time must not be so long as to interfere with speech intelligibility. Early estimates showed that reverberation times of approximately 0.6 s are fine as a compromise. However, such short reverberation times do not give speech sufficient *fullness* and do not correspond to what we expect when we listen to speech in large rooms. It is necessary to satisfy both the auditory and visual impressions.

The subjective loudness of sound in a room is estimated using the metric *room gain.* The room gain is given in dBs and is defined as the SPL in the room, less the SPL that the sound source would give at 10 m distance, if there were no room reverberation. Typical values for room gain are in the range 0 to 10 dB.

We are quite sensitive to the loudness of sounds; an orchestra of 100 persons sounds much louder than one of 50 persons, yet the difference is only approximately 3 dB. Some of this perceived increase in loudness may be due to visual influence but one must notice that psychoacoustic research shows that SPL differences of 1 dB are readily distinguished.

3.9 SOUND PRESSURE LEVEL BEHAVIOR IN ROOMS

In Chapter 2, we learned that sound in free space or open air falls off at a rate of –6 dB in SPL per doubling of distance from the sound source. In contrast, the pure spherical spreading and fall-off rate occurs only near the sound source in rooms. Further away, the energy becomes more constant as distance from the source increases because of the preponderance of reflected energy.

For continuous, wideband sounds in a reverberant space, the average sound pressure level (SPL) at various distances from the source will be approximately that given in Figure 3.12.

The region of the room in which the –6dB per distance doubling rule occurs is called the *near field*, and the region where the level is approximately constant is called the *far field* or the reverberant field.

For many musical sounds, and for speech, the situation is similar although the reverberant field for each wave packet does not become as developed as for a continuous sound. Nevertheless, from the viewpoint of estimating the signal to reverberation ratio, Figure 3.12, in principle, is correct.

The strength of the sound energy, the intensity, and the SPL in the room are dependent on the power of the sound source, its physical and radiation directional

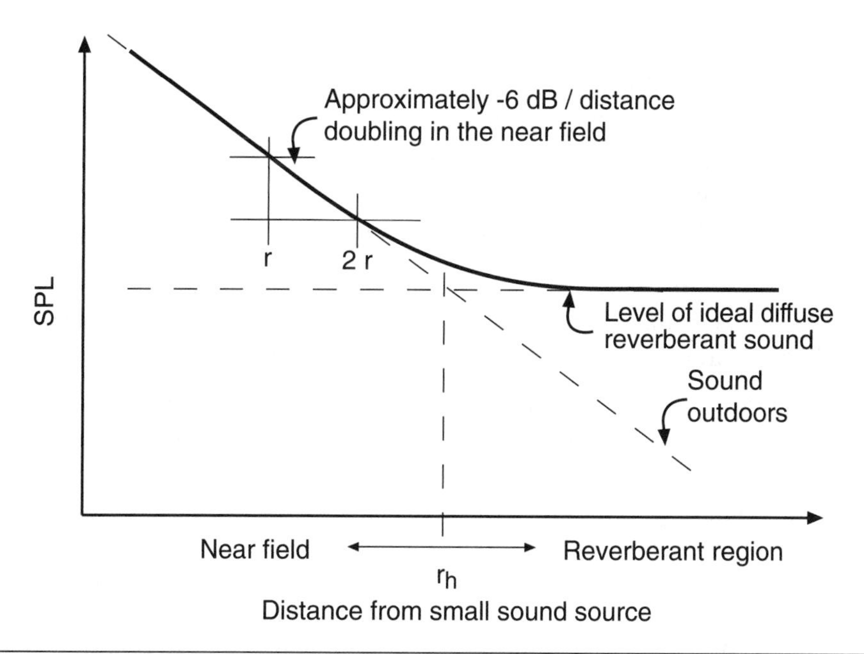

Figure 3.12 Reduction of sound with distance outdoors and within a room. Outdoors the SPL drops continuously with distance whereas the reverberation of the room tends to keep SPL relatively constant beyond a certain distance.

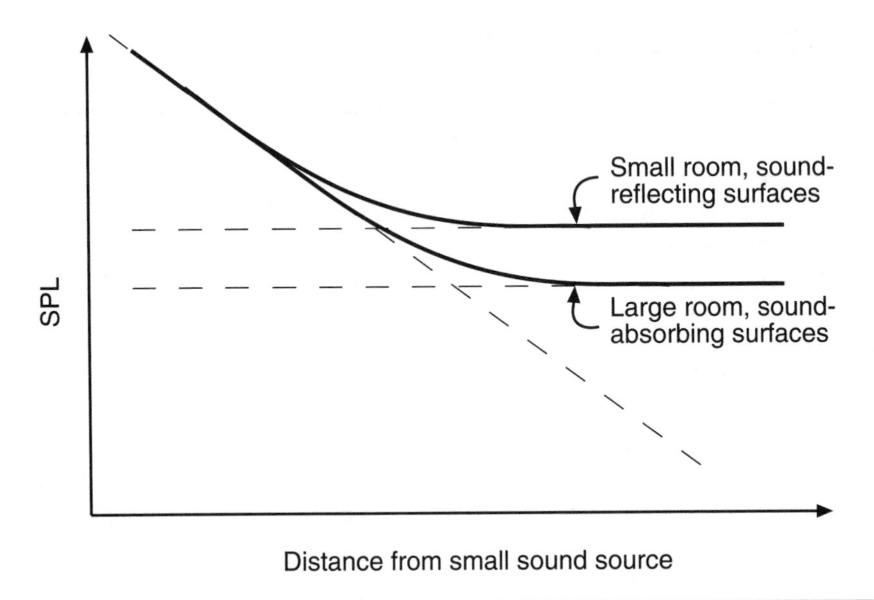

Figure 3.13 The SPL of the reverberant sound will depend on both reverberation time and room volume, but typically the level of the reverberant sound will drop if there are many sound-absorbing surfaces and if the room is large.

characteristics, the size of the room, the sound-absorption characteristics of the floor, wall, and ceiling surfaces, furnishings, people, and air. The reverberant sound pressure level, *SPLR*, is given by the sound power level of the source *PWL* and the total sound absorption *A* of the room (in metric sabins):

$$SPLR \approx PWL + 10 \log (4/A)$$

Since the reverberant sound energy is not ideally distributed throughout the room, the *SPLR* will continue to drop, but at a low rate, away from the sound source.

Two important ideas that we can glean from the above analysis are:

1. Choir and group instrumental rehearsal rooms should have ceilings at least three meters high because SPLs will be annoyingly loud if the ceiling is lower.
2. Sound-absorbing material placed more than two meters from an operator and machine tool will not reduce the noise of the tool noticeably for that operator because the operator is in the near field and the absorption will only be effective in the reverberant field (see Figure 3.13).

4

SOUND-ABSORBING MATERIALS

4.1 INTRODUCTION

Sound absorbers are used to reduce the sound levels in noisy rooms and to optimize the acoustic conditions in rooms for voice and music. Absorbers are also used to reduce sound transmission through ducts and silencers.

Sound-absorbing materials and constructions are grouped into resonant and nonresonant absorbers. Nonresonant absorbers are usually made of porous materials and, therefore, are frequently called porous absorbers. This group includes mineral or glass wool, textiles, and carpets. The resonant absorbers are made using resonant acoustic or vibratory systems such as air columns and volumes, membranes, and plates. Other absorbers are audience and performers (mostly because of clothing), as well as such sound absorptive surfaces as drapes, upholstered chairs and other furniture, walls, ceilings, carpets, and floors.

4.2 ABSORPTION COEFFICIENT AND ABSORPTION AREA

Materials are characterized by their sound-absorption coefficient, α. An α value of one represents 100% absorption of the incoming sound waves and a value of zero for the total reflection of sound. With most materials this varies with frequency, many being more sound-absorbing at high frequencies than at low frequencies. By combining materials in a room, one can provide sound-absorbing characteristics that are uniform with respect to frequency. This is useful in radio and television studios and control rooms.

In most cases one can neglect the sound power transmitted through the absorber. However, there are two special cases that need consideration: open windows and thin, sound-transparent drapes or screens, for example. In the first case, one does not have absorption in the sense that sound power is escaping from the room rather than being converted to heat. At high frequencies the open window will have an absorption coefficient close to 1.

One can also characterize the sound-absorption properties of an absorber by its absorption area, A, as measured or calculated. For a large surface, the absorption area is the product of the surface area and the absorption coefficient. Note that the absorption area may be given in either metric sabins or in sabins that are given in square ft. For example, 10 m^2 of a material having $\alpha = 0.98$ has an absorption area of 9.8 metric sabins but about 106 sabins since 10 m^2 is approximately 108 square ft.

In some cases it is more relevant to express the absorption as absorption area rather than absorption coefficient. For example, this applies to resonant absorbers and to absorbers for which the geometrical surface can be hard to define. This is typical of audience and performers as well as other types of natural absorbers.

The measurement of the absorption coefficient is accomplished by using a reverberation chamber that is a large empty room in which all surfaces are hard so that they will absorb a minimum of acoustic energy. Typically, the room surfaces will be made of painted concrete. The reverberation time (*RT*) is measured for the desired range of frequencies with and without the test sample, and Sabine's equation is used to determine the difference in sound absorption between the two cases from which the absorption coefficient is calculated using Sabine's equation (see Equation 3.1 or 3.2).

Indicative examples of typical sound-absorption properties of many materials and constructions may be found in Table 4.1. The noise reduction coefficient (NRC) shown is the average of the sound-absorption coefficient in the 0.25, .5, 1, and 2 kHz octave bands rounded to the nearest multiple of 0.05.

Note that the absorption coefficient, α, is the same for both metric and feet calculations because it is the ratio of absorbed to incident sound. The figures for the area for a particular absorbing surface vary and thus the equations for the effect of the absorption must be different, as discussed in Chapter 3.

4.3 POROUS ABSORBERS

The conversion of acoustic to thermal energy in porous absorbers is due to the viscous behavior of air flowing in the canals, pores, and air pockets of the material. The porosity of a material is decisive for its sound-absorption properties. Commercial products for sound absorption such as glass wool may have a porosity larger than 0.95. As an example a 10 cm (4 in) thick sheet of glass fiber will typically absorb nearly 98% of the sound that is incident in most of the speech frequency range. Note that when a thick sound-absorbing sheet is suspended freely, for example when hanging vertically from a ceiling, both sides will, of course, contribute to the sound absorption.

Table 4.1 Examples of typical measured sound-absorption coefficients for various materials and constructions

	Octave-band center frequencies						
	125 Hz	250 Hz	500 Hz	1 kHz	2 kHz	4 kHz	NRC
Walls							
Concrete block, grout-filled, and unpainted	0.01	0.02	0.03	0.04	0.05	0.07	0.05
Concrete block, grout-filled, and painted	0.01	0.01	0.02	0.02	0.02	0.03	0.00
Poured concrete, unpainted	0.01	0.02	0.04	0.06	0.08	0.10	0.05
Normal, unpainted concrete block	0.36	0.44	0.31	0.29	0.34	0.21	0.30
Painted concrete block	0.10	0.05	0.07	0.09	0.08	0.05	0.06
6 mm heavy plate glass	0.18	0.06	0.04	0.03	0.02	0.02	0.05
2.4 mm regular window glass	0.36	0.25	0.18	0.12	0.07	0.04	0.15
12 mm plasterboard, wood studs spaced 0.4 m	0.29	0.10	0.05	0.04	0.12	0.09	0.05
Plasterboard, 1.6 m × 0.8 m panels on 62 mm studs spaced 0.4 m (voids filled with glass fiber)	0.55	0.14	0.08	0.04	0.10	0.10	0.10
Plasterboard, 1.6 m × 0.8 m panels on 128 mm studs spaced 0.4 m (voids filled with glass fiber)							
Marble and glazed brick	0.01	0.01	0.01	0.01	0.02	0.02	0.00
Painted plaster on concrete	0.01	0.01	0.01	0.03	0.04	0.05	0.05
Plaster on concrete block or 25 mm on metal studs	0.12	0.09	0.07	0.05	0.05	0.04	0.05
Plaster 16 mm on metal studs	0.14	0.10	0.06	0.05	0.04	0.03	0.05
6 mm wood over air space	0.42	0.21	0.10	0.08	0.06	0.06	0.10
9-10 mm wood over air space	0.28	0.22	0.17	0.09	0.08	0.06	0.05
25 mm wood over air space	0.19	0.14	0.09	0.06	0.06	0.05	0.10

Table 4.1 *(Continues)*

	Octave-band center frequencies						
	125 Hz	**250 Hz**	**500 Hz**	**1 kHz**	**2 kHz**	**4 kHz**	**NRC**
Venetian blinds, open	0.06	0.05	0.07	0.15	0.13	0.17	0.10
Velour drapes, 280 g/m^2	0.03	0.04	0.11	0.17	0.24	0.35	0.15
Fabric curtains, 400 /m^2 hung 1/2 length	0.07	0.31	0.49	0.75	0.70	0.60	0.55
Same, but 720 g/m^2 hung 1/2 length	0.14	0.68	0.35	0.83	0.49	0.76	0.55
Shredded wood panels, 50 mm on concrete	0.15	0.26	0.62	0.94	0.64	0.92	0.60
12 mm wood panels, holes 5 mm, 11% open area over void	0.37	0.41	0.63	0.85	0.96	0.92	0.70
Same with 50 mm glass fiber in void	0.40	0.90	0.80	0.50	0.40	0.30	0.65
Floors							
Terrazo	0.01	0.01	0.02	0.02	0.02	0.02	0.00
Glazed marble	0.01	0.01	0.01	0.01	0.02	0.02	0.00
Linoleum, vinyl, or neoprene tile on concrete	0.02	0.03	0.03	0.03	0.03	0.02	0.05
Wood on joists	0.10	0.07	0.06	0.06	0.06	0.06	0.05
Wood on concrete	0.04	0.07	0.06	0.06	0.06	0.05	0.05
Carpet on concrete	0.02	0.06	0.14	0.37	0.60	0.65	0.30
Carpet on expanded (sponge) neoprene	0.08	0.24	0.57	0.69	0.71	0.73	0.55
Carpet on felt under pad	0.08	0.27	0.39	0.34	0.48	0.63	0.35
Indoor-outdoor thin carpet	0.10	0.05	0.10	0.20	0.45	0.65	0.20
Ceilings (also check floors and walls for similar material)							
Suspended 12 mm plasterboard							
Suspended 12 mm plasterboard with steel	0.29	0.10	0.05	0.04	0.07	0.09	0.05
Same with steel suspension system frame	0.15	0.10	0.05	0.04	0.07	0.09	0.14

Plaster on lath suspended with framing	0.14	0.10	0.06	0.05	0.04	0.03	0.05
Glass fiber acoustic tile suspended	0.76	0.93	0.83	0.99	0.99	0.94	0.95
Wood fiber acoustic tile suspended	0.59	0.51	0.53	0.71	0.88	0.74	0.65
Acoustic tile, 50 mm on drywall, suspended	0.08	0.29	0.75	0.98	0.93	0.76	0.75
Acoustic tile, 50 mm on drywall, suspended air-space between tile and drywall	0.38	0.60	0.78	0.80	0.70	0.75	0.75
Glass fiber, 360 g/m², suspended	0.65	0.71	0.82	0.86	0.76	0.62	0.80
Glass fiber, 1120 g/m², suspended	0.38	0.23	0.17	0.15	0.09	0.06	0.15
Same but 1120 g/m²	0.38	0.23	0.17	0.15	0.09	0.06	0.15
Luminous ceiling, typical with openings	0.07	0.11	0.20	0.32	0.60	0.85	0.30
Hanging absorbers, 450 mm, 450 mm on centers	0.07	0.20	0.40	0.52	0.60	0.67	0.45
Hanging absorbers, 450 mm, 159 mm on centers	0.10	0.29	0.62	0.72	0.93	0.98	0.85
Seats and people (on area basis, not per-person)							
Unoccupied seats, light upholstery, 20 mm	0.36	0.47	0.57	0.62	0.62	0.60	
Occupied seats, light upholstery, 20 mm	0.51	0.64	0.75	0.80	0.82	0.83	
Unoccupied seats, medium upholstery, 50 to 100 mm	0.54	0.62	0.68	0.70	0.68	0.66	
Occupied seats, medium upholstery, 50 to 100 mm	0.62	0.72	0.80	0.83	0.84	0.85	
Unoccupied seats, heavy upholstery, 200 mm+	0.70	0.76	0.81	0.84	0.84	0.81	
Occupied seats, heavy upholstery, 200 mm+	0.72	0.80	0.86	0.89	0.90	0.90	
Wood benches and pews, occupied	0.57	0.61	0.75	0.86	0.90	0.86	

Table 4.1 *(Continues)*

	Octave-band center frequencies						
	125 Hz	**250 Hz**	**500 Hz**	**1 kHz**	**2 kHz**	**4 kHz**	**NRC**
Wood benches and pews, unoccupied	0.15	0.19	0.22	0.39	0.38	0.30	
Students in tablet-arm seats	0.30	0.41	0.49	0.84	0.87	0.84	
Special material							
Organ at Boston Symphony Hall, metric Sabins measured by covering chamber opening with heavy, hard, sound-reflecting material and comparing reverberation times	41	26	19	15	11	11	
Organ at Tokyo Opera City, metric Sabins measured by comparing reverberation times before and after free-standing organ installation	65	44	35	33	32	31	
Surfaces outdoors							
Lawn grass, 50 mm high	0.11	0.26	0.60	0.69	0.92	0.99	0.70
Fresh snow	0.45	0.75	0.90	0.95	0.95	0.95	0.90
Plowed soil	0.15	0.25	0.40	0.55	0.60	0.60	0.45
Spruce trees 2.7 m	0.03	0.06	0.11	0.17	0.27	0.31	0.15
Water in a swimming pool	0.01	0.01	0.01	0.02	0.02	0.03	0.00
Openings							
Open entrances off lobbies and corridors: 0.50 to 1.00							
Air supply and return grilles: 0.15 to 0.50							
Stage-house proscenium opening: 0.25 to 0.75							
Residual absorption coefficient	0.14	0.12	0.10	0.09	0.08	0.07	

Porous materials that do not have open cells or pores cannot function as porous absorbers; an example of such a material is expanded polystyrene. If the flow is blocked by paint or other surface coverings, the absorption coefficient will usually drop considerably. Because of the elasticity of the material, there may still be some absorption.

It is important to differentiate between the sound-absorption and the heat-insulation properties of porous materials. Porous materials may be effective thermal insulators and still have poor sound-absorption properties. Expanded polystyrene and many other porous plastic materials with closed pores have poor sound-absorption properties.

Figure 4.1 shows how the sound-absorption coefficient for a porous-absorber sheet varies as a function of sheet thickness when the sheet is mounted immediately in front of a hard surface. If the sheet is thin, there will be some sound remaining in the sheet that will be reflected as it meets the hard surface. Since the sound absorption of a porous sheet depends on the possibility for air flow through

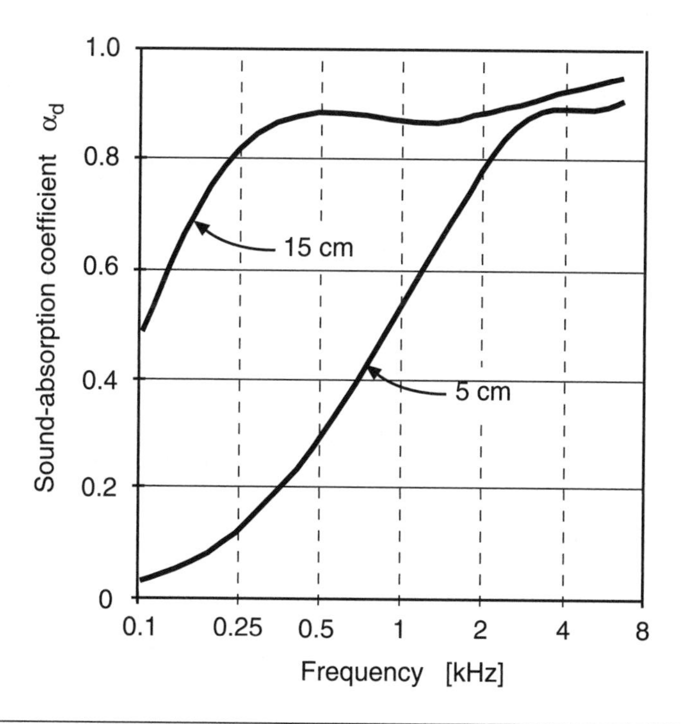

Figure 4.1 Typical calculated sound-absorption characteristics of a porous sound-absorbing material mounted directly against a hard surface as a function of frequency. Characteristics are shown for two material thicknesses.

the material, it is not optimal to position a material characterized by easy flow flush to a hard surface.

To increase the sound absorption of porous sound absorbers at low frequencies, one should use a suitable combination of material flow resistance, thickness, and air space distance. Nearly any porous sound absorber will benefit from being mounted with some air space between itself and a hard wall. The data shown in Figure 4.2 shows the beneficial influence of such air space.

One might be tempted to use textile draperies and similar materials for sound absorption, but it must be noted that thin materials, including cloth, should be hung in deep folds in front of a wall, otherwise their sound absorption will vary considerably with frequency.

4.3.1 Disadvantages of Porous Sound Absorbers

Porous absorbers are easy to use and offer effective sound-absorbing properties in many situations, but they also have a number of disadvantages and can create problems. Some examples of the problems are their sensitivity to mechanical

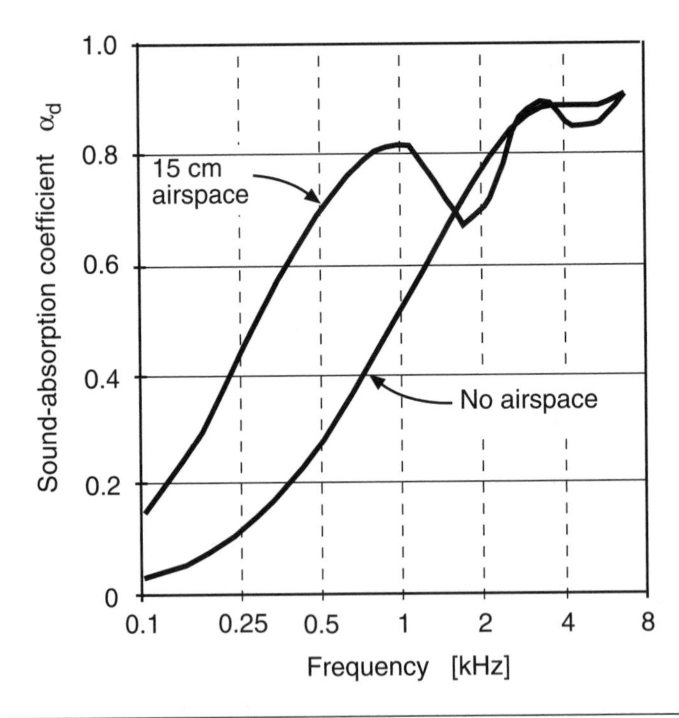

Figure 4.2 Typical calculated sound-absorption characteristics of a 5 cm porous sound-absorbing material mounted directly against a hard surface and mounted with 15 cm of air space as a function of frequency.

damage, they do collect dust, particularly close to air exhausts and intakes and, when touched, they may release hazardous fibers. Other important disadvantages are that they cannot be painted because the pore openings will be closed and they cannot be washed because pores will be clogged by water.

To avoid some or all of these disadvantages and problems, it is possible to cover the surface of a porous material with various sheets made of glass fiber cloth, or perhaps plastic or metal foils. These covers may have varying degrees of limpness, mass, and compliance. Any protective sheet must be extremely lightweight, as indicated by the data shown in Figure 4.3.

The apparent surface mass of the limp sheet will increase if the sheet touches the absorber. The protective foils or sheets will also help to avoid material fibers, which may be a health hazard, from contaminating the surrounding air. For the protection to be most effective, the sheets must also be mounted away from the porous absorber. This is difficult to achieve in practice because most surface protected commercial sound-absorptive materials have the foil attached to the skeleton of the material.

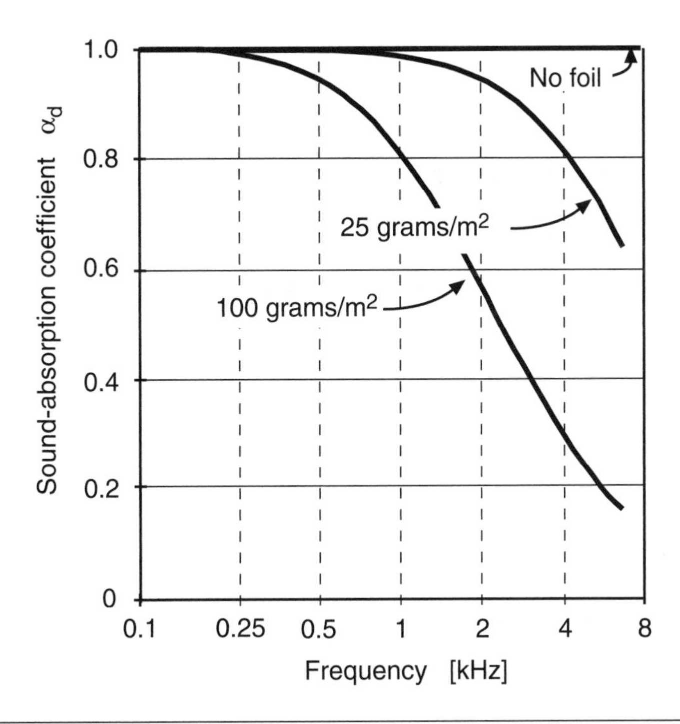

Figure 4.3 Calculated sound-absorption characteristics of a perfect porous absorber protected by limp foils having two different mass per surface area properties.

Highly perforated boards made of gypsum or plywood, for example, may also be used to give mechanical protection to porous absorbers. They should then have a maximum distance between holes of 5 cm (2 in) and be perforated over at least 20% of the surface area. For fewer and/or smaller holes, the boards will exhibit resonance-absorber characteristics.

4.4 RESONANT ABSORBERS

The sound absorption by resonance absorbers relies on the losses in their acoustical and mechanical constructions that are set in motion by sound. The most common types of resonance absorbers are Helmholtz, membrane, and slit absorbers. Thin panels, walls, and windows can also act as resonant sound absorbers. One can combine porous and resonant-absorber action in a material. Porous materials can be designed to hold resonant elements and vice versa.

Resonant absorbers are primarily used for absorbing sound with frequencies below 400 Hz, which means that they are an ideal complement to porous

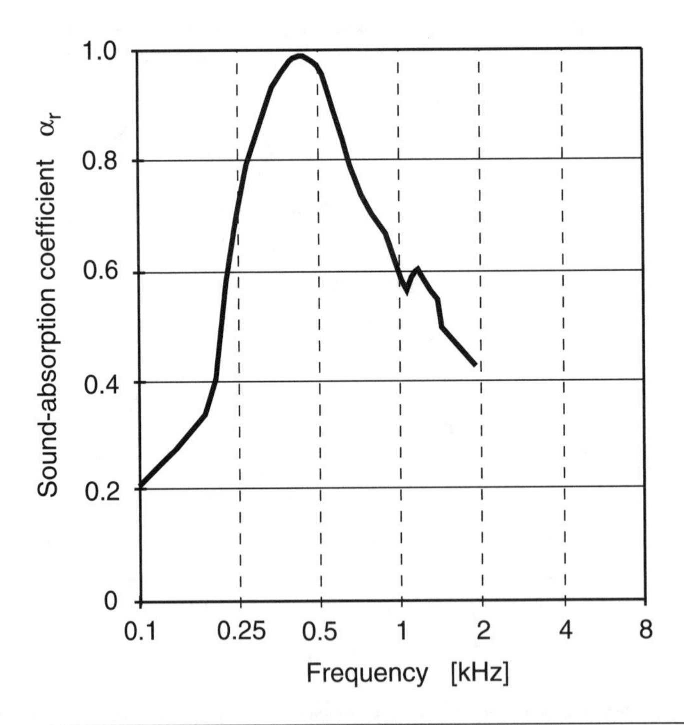

Figure 4.4 Measured sound-absorption characteristics of a membrane absorber using a plastic foil having a surface mass of 125 g/m^2 with a 90 mm air space (filled with a glass wool) in front of a hard wall.

absorbers. In isolated cases, one can design resonance absorbers to cover a wide frequency range, including high frequencies. One example of such a construction that has smooth, wide-range sound absorption, is a wall of air bricks inside which is an air space with a porous absorber.

4.4.1 Membrane-type Absorbers

A membrane absorber can be designed to use a flexible membrane, impervious cloth, or a plastic foil mounted so that a closed air space is formed between the membrane and the underlying wall. This air space will act as a cushion. The mass of the membrane and the spring action of the air cushion result in a resonant mechanical system that is the same as that of a drum. Figure 4.4 shows the typical resonant characteristics of a membrane sound absorber. Other common examples of membrane absorbers are windows, wooden floors on joists, and wooden wall paneling on battens.

The degree of resonance will depend on losses in the membrane and in the air space that can be filled with a porous absorber. Note that the addition of a porous absorber will affect the resonance frequency and the bandwidth of the absorption peak because of its added damping effect; the less damping, the higher the relative sound-absorption peak. Resonance absorbers that are effectively absorptive over an octave's frequency range are the most useful. Since it is difficult to estimate the damping and the resulting absorption coefficient in advance, the absorption characteristics are best found by measurement (see Reference 3.1).

The resonance frequency can be calculated from the formula:

$$f \approx \frac{60}{\sqrt{md}} \quad \text{(metric units)}$$

where m is the mass per unit area of the membrane (in kg/m^2) and d is the depth of the air space (in meters).

Typically, the resonance frequency will be chosen to be in the 50 to 400 Hz range when one uses panels that have a thickness of approximately a centimeter and air spaces that have a depth of a few centimeters. In these cases, one usually finds that the maximum sound-absorption coefficient is 0.8 or lower.

If one wants high absorption over a small frequency range, one must use a thin panel and a large air-space distance. Thin panels, however, frequently feature multiple absorption peaks due to air-space room resonances.

4.4.2 Helmholtz Resonators

An open bottle, an urn, or a guitar body are all examples of a simple Helmholtz resonator. The resonator can be free-standing or inserted into a wall with its opening facing the room.

The air in the bottle's neck constriction will act as the mass, and the air in the flask body acts as the spring. The Helmholtz resonator has intrinsically minor damping so its proper damping is an important factor in design. The absorption maximum will depend on placement and damping. One can adjust the damping by adding a porous absorber inside the flask body or in the throat constriction. The damping adjust the range over which useful sound absorption occurs (see Reference 3.1).

The frequency at which the absorption maximum occurs is given by the geometry of the flask's neck and body:

$$f \approx 55\sqrt{\frac{S}{lV}} \quad \text{(Metric units)}$$

where S is the neck's cross sectional area (in m^2), l its length (in m), and V the air volume of the body (in m^3).

By using a number of resonators that have different resonance frequencies, one can achieve more wideband absorption than by using a single resonator.

The absorption of a Helmholtz absorber is best described by the idea of an absorption area, because the equivalent absorption can be much larger than the area of the throat. Useful sound absorption over a large frequency range requires a large enclosed air space. The shape of the air space is quite unimportant as long as the air space has small dimensions compared to the wavelength. Figure 4.5 shows the resonant characteristics of air bricks used as Helmholtz resonators. Note that the measured sound-absorption coefficient is higher than unity, indicating that the sound-absorption area is larger than the physical dimensions of the sample.

An advantage of Helmholtz resonators in room acoustics is that they can be used to control the decay of individual low-frequency room resonances. Such low-frequency resonators are sometimes called bass traps.

4.4.3 Resonator Panels

A sound absorber similar in action to the Helmholtz resonator is the resonator panel that consists of an air space and a panel. The panel needs to have perforations such as slits or circular holes. One can regard such panels as a large number of individual Helmholtz resonators next to one another. The resonance frequency of such a panel can be determined similarly to that of a Helmholtz resonator (see Reference 3.1).

As mentioned, the holes do not need to be cylindric; slits and other hole shapes can be used as long as the maximum crosswise dimensions are much smaller than the wavelength. By damping the resonance using materials such as cloth or mineral wool, one can obtain quite wide bandwidth, typically more than one third-octave band wide. The sound-absorption coefficient at the resonance

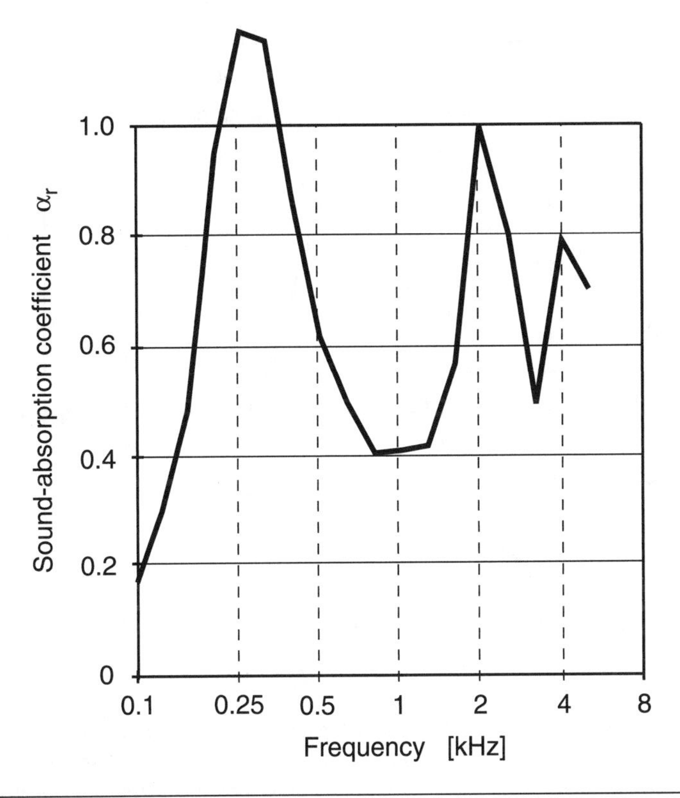

Figure 4.5 Measured sound-absorption characteristics of air bricks used to form a Helmholtz resonator. Neck length 70 mm with 50 mm air space in front of a hard wall.

frequency is typically high, often close to unity. There are many commercial resonator panel designs. The panel normally will be made of gypsum or plywood and designed to be a part of the ceiling of a room. Figure 4.6 shows the measured sound-absorption characteristics of a resonator panel consisting of a 1 mm perforated steel sheet (square pattern 5 mm diameter holes at 20 mm distance) with a 95 mm air space in front of a hard wall. A 15 mm porous sound absorber was inserted into the air space. The graph shows the results for two different positions of the absorber. The flow resistance of the porous material helps increase the sound absorption when it is mounted close to the holes.

4.5 ADJUSTABLE SOUND ABSORPTION

Ideally, one wants to be able to adjust the sound absorption of the additional absorbing areas in a room to that which is required in the specific case because

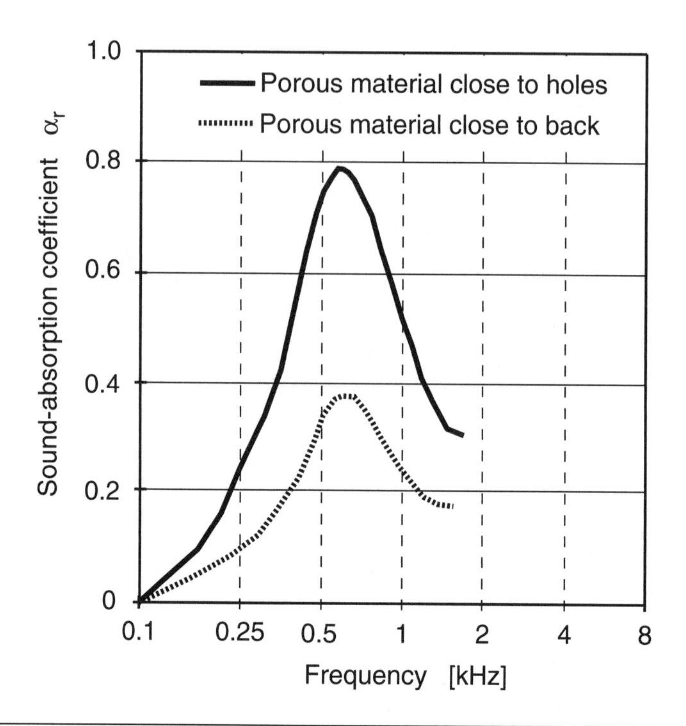

Figure 4.6 Measured sound-absorption characteristics of a resonator panel consisting of a 1 mm perforated steel sheet (square pattern 5 mm diameter holes at 20 mm distance) with a 95 mm air space in front of a hard wall. A 15 mm porous sound absorber was inserted into the air space. The graph shows the results for two positions of the porous material.

the natural absorption in a room varies, for example, depending on the number of people present, among other factors. As mentioned previously, various uses require different *RT*s. Using variable absorbers, it may be possible to adjust the *RT* to the desired value irrespective of the number of people in the audience. Some examples of the construction of such variable absorbers are shown in Figure 4.7.

4.6 SOUND ABSORPTION BY AUDIENCE

The sound absorption by the congregants' clothing is often dominating at mid-frequencies in enclosed performance spaces such as synagogues, churches, and mosques. In large rooms (typically having volumes more than 10,000 m^3 or 350,000 ft^3), the sound absorption induced by the damping properties of air will be so large as to limit the maximum *RT* that can be obtained in the room.

Due to the sound absorption by worshippers, if one wants a certain *RT* (determined by the air volume of the room), one will need a certain volume per seating

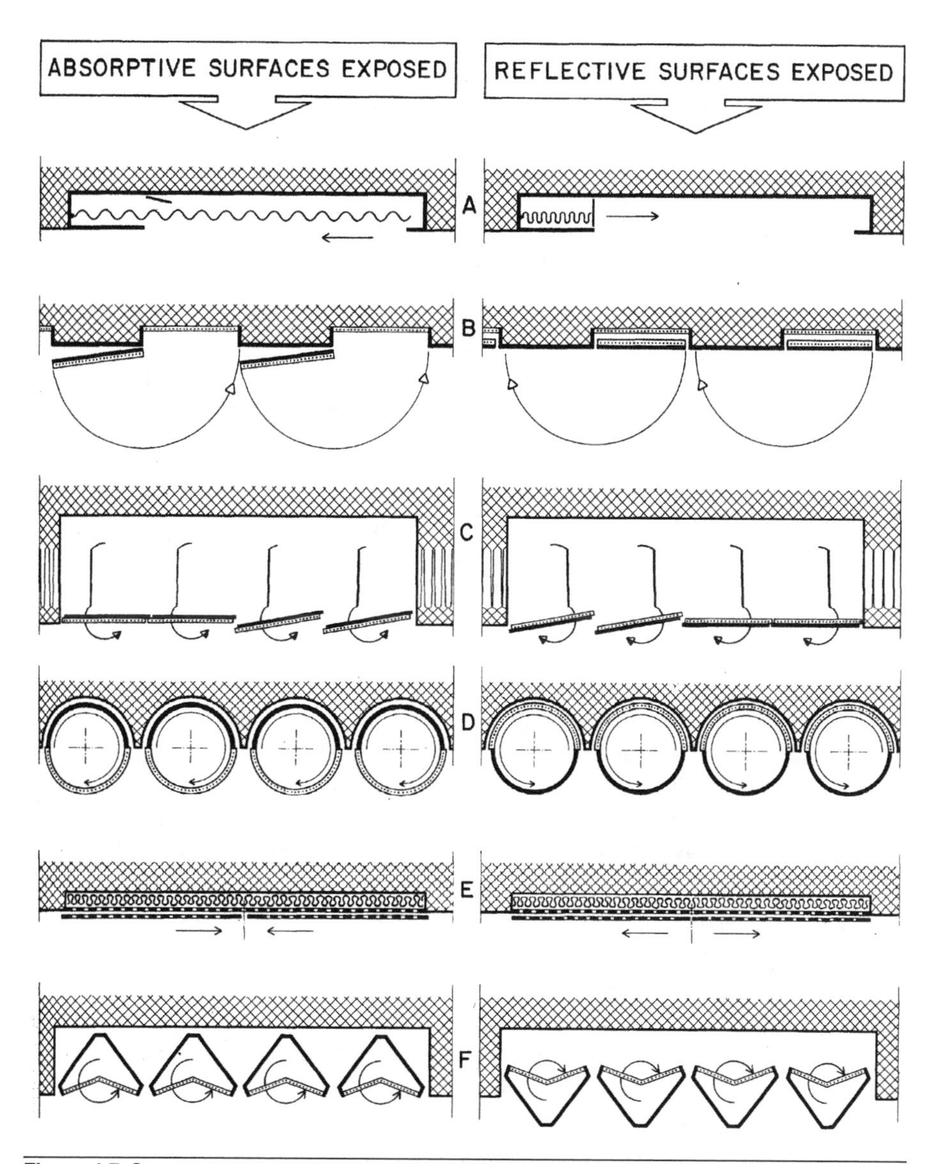

Figure 4.7 Some common designs of variable sound absorbers. (a) Retractable curtain, (b) hinged panels, (c) rotable panels, (d) rotable cylinders, (e) sliding perforated panels, and (f) rotable triangular elements (see Reference 4.4).

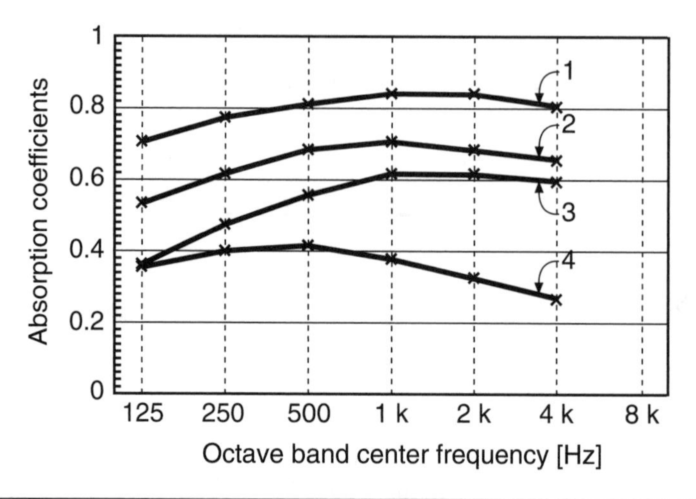

Figure 4.8 Unoccupied chair sound-absorption coefficients from four groups of concert halls according to degree of seat upholstering (see Reference 4.2). (1) Heavily upholstered seats, (2) medium upholstered seats, (3) lightly upholstered seats, and (4) extra lightly upholstered seats.

area. The worshipper absorption depends, of course, on the number of persons per unit area and the upholstering of the chairs or pews. The audience absorption also depends on the diffusitivity of the reverberant sound, thus, the design of large spaces to obtain a particular *RT* is not straightforward. Figure 4.8 shows some examples of measured audience sound-absorption characteristics in concert halls.

Another problem of worship acoustics is to design seating that has approximately the same sound absorption both empty and occupied. This design has to be empirical and can be quite expensive. Ideally, at least 25 chairs (or 5 pews with cushions) are usually necessary to have a reasonable approximation of the sound field around the worshippers for acoustic testing in a reverberation chamber.

It must be emphasized that the sound absorption due to the worshippers' presence in the room should not be confused with the attenuation due to interference as sound propagates along the audience surface.

In the post-World War II era, several concert halls and worship spaces were designed with the reverberation times (*RT*s) using Sabine's formula (Equation 3.2), but opened with measured reverberation times considerably lower than designed. The most prominent case is London's Royal Festival Hall. Dr. Leo L. Beranek, MIT professor and president of a Cambridge, Massachusetts, architectural acoustics consulting firm, faced this problem while assisting architects Zev and Yacov Rechter with the design of Tel Aviv's Mann Auditorium.

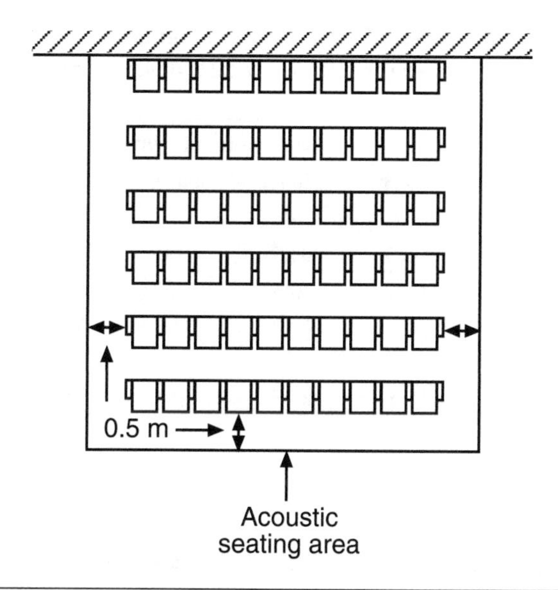

Figure 4.9 The *acoustical* audience area is the area beneath chairs plus areas of strips 0.5 m (ca. 1.5 ft) wide around audience seating blocks except for sides at balcony rails or walls (see Reference 4.2).

At the time, acoustical consultants were calculating audience sound absorption on a per person basis, but the seat spacing in modern halls and some worship spaces was much more generous than in older halls, for example, Symphony Hall in Boston. Beranek determined audience and empty-seat sound absorption is best determined on a per-area, not on a per-person basis, and the Mann Auditorium then came close to predicted values when opened in 1957.

Since audience absorption (clothing being the major part) is the largest share of the absorption in performing spaces and that of the congregation in worship spaces, Beranek's work was of extreme importance to worship space acoustics as well as concert halls and opera houses (see Reference 4.3).

Because of the sound-field properties at the edges of the seating area, it is necessary to compensate for the sound that enters from the sides. The entire *acoustical* audience area is the area beneath chairs plus areas of strips 0.5 m (ca. 1.5 ft) wide around audience seating blocks, except for sides at balcony rails or walls, as shown in Figure 4.9.

The sound absorption of pew cushions is influenced by the presence of the pews and is largely independent of the pew sample size (see Reference 4.1). For individuals seated a few meters (yards) apart, the absorption area is still calculated on a per-person basis. Some values are shown in Table 4.2 (see Reference 4.5).

Table 4.2 Examples of measured sound-absorption areas in metric sabins for individuals. Values measured for each individual having 6 m² of surrounding floor area (see Reference 4.5).

	Octave-band center frequencies					
	125 Hz	**250 Hz**	**500 Hz**	**1 kHz**	**2 kHz**	**4 kHz**
Individuals						
Man in suit, standing	0.15	0.25	0.60	0.95	1.15	1.15
Man in suit, sitting	0.15	0.25	0.55	0.80	0.90	0.90
Woman in summer dress, standing	0.05	0.10	0.25	0.40	0.60	0.75
Woman in summer dress, sitting	0.05	0.10	0.10	0.35	0.45	0.65

Note: All values are in metric sabins

4.7 RESIDUAL SOUND-ABSORPTION AREA

Any room will have sound absorption that is difficult to estimate, including the absorption by objects, lighting fixtures, ventilation openings, cracks around doors in the room, among others. Such absorption can be accounted for approximately by a residual sound-absorption coefficient (see Reference 4.2). In the rooms investigated, this α was approximately 0.1 as averaged over the entire surface area of the room.

5

METRICS FOR ROOM ACOUSTICS

5.1 INTRODUCTION

It would be advantageous if it were possible to specify the acoustic quality of a room by some simple numbers or indicators when writing building specifications or for quality control. It turns out that this is difficult because of the lack of sufficient knowledge regarding the properties of human hearing.

For example, a room may show large numerical differences between the current quality metrics measured at various positions in the room whereas the subjective experience of listening in the room does not correlate to these differences. It is, in particular, the fine structure of the reverberation that is difficult to measure. The metrics require considerable experience and expertise to be applied correctly. Often, listening and subjective judgment by experts is the only way to assess the room acoustics quality. Hearing is an important measurement device.

In laboratory experiments, the influence of visual cues can be removed, but the sound field must be approximated. Doing experiments in real auditoria, it is difficult to remove the visual cues and other cognitive factors.

5.2 IMPULSE RESPONSE

In a real room there will be many reflected sound-field components that together become reverberation. The room impulse response at some point in the room is defined as the relationship between a short sound pulse at 1 m distance from an omnidirectional source (electroacoustic transducer, spark, etc.) and the actual

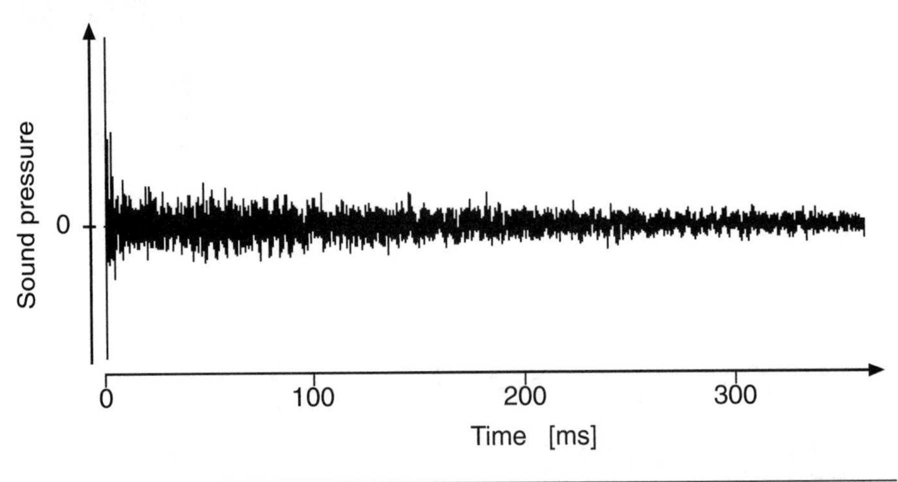

Figure 5.1 A room-impulse response measured close to the loudspeaker in St. Jacobi, a large church in Hamburg, Germany.

sound pressure response at the measurement point. The impulse response is a function of time, $h(t)$. It corresponds to the sound reflection time pattern discussed in Chapter 3. By using directional microphones, other types of room-impulse responses may be measured and used for room acoustic analysis. Today, most metrics of the acoustical quality of a room are based on an initial measurement of the room's impulse response $h(t)$ from which the various metrics are derived from the use of equipment or software. An example of a measured room-impulse response for a large church is shown in Figure 5.1.

5.3 REVERBERATION TIME

From the viewpoint of subjective importance, after loudness, reverberation time (*RT*) is generally the property of a room that is the most important for the subjective impression of room acoustics. It was the first room acoustics property to be investigated on a truly scientific basis.

According to the ISO3382 standard, the measured reverberation time is obtained by extrapolation to 60 dB of a line fitted to the reverberation curve (see Figure 3.11). In practice, however, it is often difficult to achieve a signal-to-noise ratio that allows the full range of 60 dB to be used; instead the curve is extrapolated from −5 dB to −35 dB of decay, for example. The reverberation time is then best written as $RT_{-5,-35}$. In spite of the fact that it is seldom actually being measured over a 60 dB level drop, many people still write T_{60} in practice. It is difficult to obtain a good range when measuring reverberation time in large venues.

Unless otherwise noted, the room-impulse responses are to be measured using an omnidirectional microphone. The standard also requires the measurements to be completed using at least two sending positions and five receiving positions spaced in a specified way on the stage and in the audience seating area.

It is difficult to measure reverberation time with worshippers and religious leaders present even during a pause in services. The measurements take some time, and the worshippers seldom want to be part of an experiment. Considerable experience is required to judge the difference in reverberation characteristics between a room that is occupied and one that is unoccupied.

Measurement of reverberation time can be quite tricky and require analysis of the decay history of the sound pressure. It may then be necessary to adjust the levels over which the reverberation is measured and the time extrapolated. This applies when the reverberation curves feature dual slopes, for example, in the various subspaces in a church having a cruciform floor plan or a cathedral having many subspaces. Figure 5.2 illustrates the behavior of reverberation when two acoustically coupled spaces have significantly different reverberation characteristics.

Many people have come to like the acoustics characterized by dual reverberation slopes. That is likely due to the fact that when listening to music in the home, over stereo or multi-loudspeaker systems, the reverberation in the recording is mixed with the reverberation of the listening space. Dual-slope decays also make it possible to create listening situations that have excellent clarity coupled with long reverberation times and much ambience.

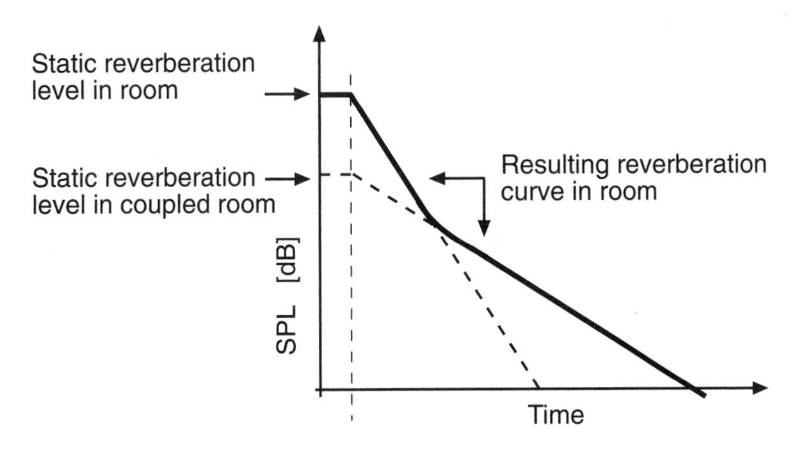

Figure 5.2 Example of a dual-slope reverberation curve for a room that is acoustically somewhat coupled to another room having a longer RT.

5.4 EARLY DECAY TIMES

The initial part of the room-impulse response is determined by the early reflections. The metric early decay time (*EDT*) that is based on the first 10 dB of reverberation process measures this decay so $EDT = RT_{0,-10}$. The *EDT* values for some octave bands are used in the definition of the Brilliance metric.

A slightly different metric is the reverberation time for the first 160 ms of the room-impulse response, RT_{160}. It has been shown to correlate better with the subjective feeling of reverberation time for music.

5.5 CLARITY—EARLY-TO-REVERBERANT RATIO

The ratio of early-to-reflected sound, expressed in dB, between a summation of all sound received from the initial sound to the reflection arriving at 50 ms (for speech and some music) or to the reflection arriving at 80 ms (for most music), is a strong measure of clarity, in general, and intelligibility of speech. It is typically evaluated using octave-band analysis.

Clarity, C_{50} for speech, is defined as:

$$C_{50} = 10\log\left(\frac{\int_0^{50} h^2(t)dt}{\int_{50}^{\infty} h^2(t)dt}\right) \quad [\text{dB}]$$

whereas Clarity, C_{80} for music, is defined as:

$$C_{80} = 10\log\left(\frac{\int_0^{80} h^2(t)dt}{\int_{80}^{\infty} h^2(t)dt}\right) \quad [\text{dB}]$$

Clarity is difficult to predict from drawings and usually requires computer simulation or scale model testing for accurate advance predictions and are easily measured with systems, including the Crown TEF system, DRA Labs MLSSA system, and many others. Clarity for speech correlates well with the Articulation Index and intelligibility testing. It is at least as important as any other metric in room design that we discuss later. For speech, a weighted average of the 500 Hz to 4 kHz octave bands is often employed, whereas for music, an unweighted of the 125 Hz to 4 kHz octave band is used. Typically, clarity will be in the range of –5 to 5 dB.

5.6 INITIAL TIME-DELAY GAP

Time (in milliseconds) between arrival of direct sound and the first useful reflected sound at a listener's ears is known as initial time-delay gap (*ITG*). It is the measure (along with the Clarity) of the subjective *intimacy* of a room that was the first quality of a room uncovered by Beranek in his search for other qualities beyond reverberation time (see Reference 5.1).

5.7 SPEECH INTELLIGIBILITY AND ARTICULATION

Correctly understood speech is crucial in many settings, such as classroom teaching, warning systems, flight control systems, among others. In these venues, speech is usually *contaminated* by noise, echo, and reverberation.

Depending on how speech intelligibility is defined and measured, one can obtain varying measurement results. The intelligibility can be estimated, for example, by the percentage of correctly perceived phonemes, single-syllable nonsense words, single-syllable phonetically balanced words, and sentences. The choice of word lists, listener ability (hearing threshold and intellectual capacity), and presentation techniques all influence the intelligibility test results. In addition, the talker's voice, including spectrum and pronunciation, will influence the test results.

The percentage of correctly understood items is often called articulation, and, therefore, it is usually necessary to specify which type of word material or sounds were used in the test, by specifying sentence articulation. To assess speech intelligibility, it is necessary to use listeners; consequently, the results will show substantial statistical uncertainty.

5.8 SPEECH INTELLIGIBILITY METRICS

Due to the inconvenience and the cost in determining speech intelligibility using actual listeners, many metrics have been developed to estimate speech intelligibility. These metrics are only approximations, therefore, one should always be cautious when using speech intelligibility metrics, particularly within the field of room acoustics. Many metrics were developed for noise-infused, single-channel, analog telephone communication situations and may not sufficiently measure the influence of room acoustic conditions such as the presence of masking reverberation.

Articulation Index (*AI*) is a metric designed in the 1940s to estimate the quality of radio telephony and was intended to be used for single-channel voice communication in the presence of noise. It is estimated by listening tests, using the number of correctly understood voiced and unvoiced sounds according to a special formula. One can also use a table or nomogram by which the *AI* can be estimated from the signal-to-noise ratio. A similar index, the Speech Intelligibility Index, defined by an ANSI standard is also used and is more modern.

Articulation Loss of Consonants (*ALCons*) is still a popular metric among designers of speech reinforcement systems. It can be measured using test listeners

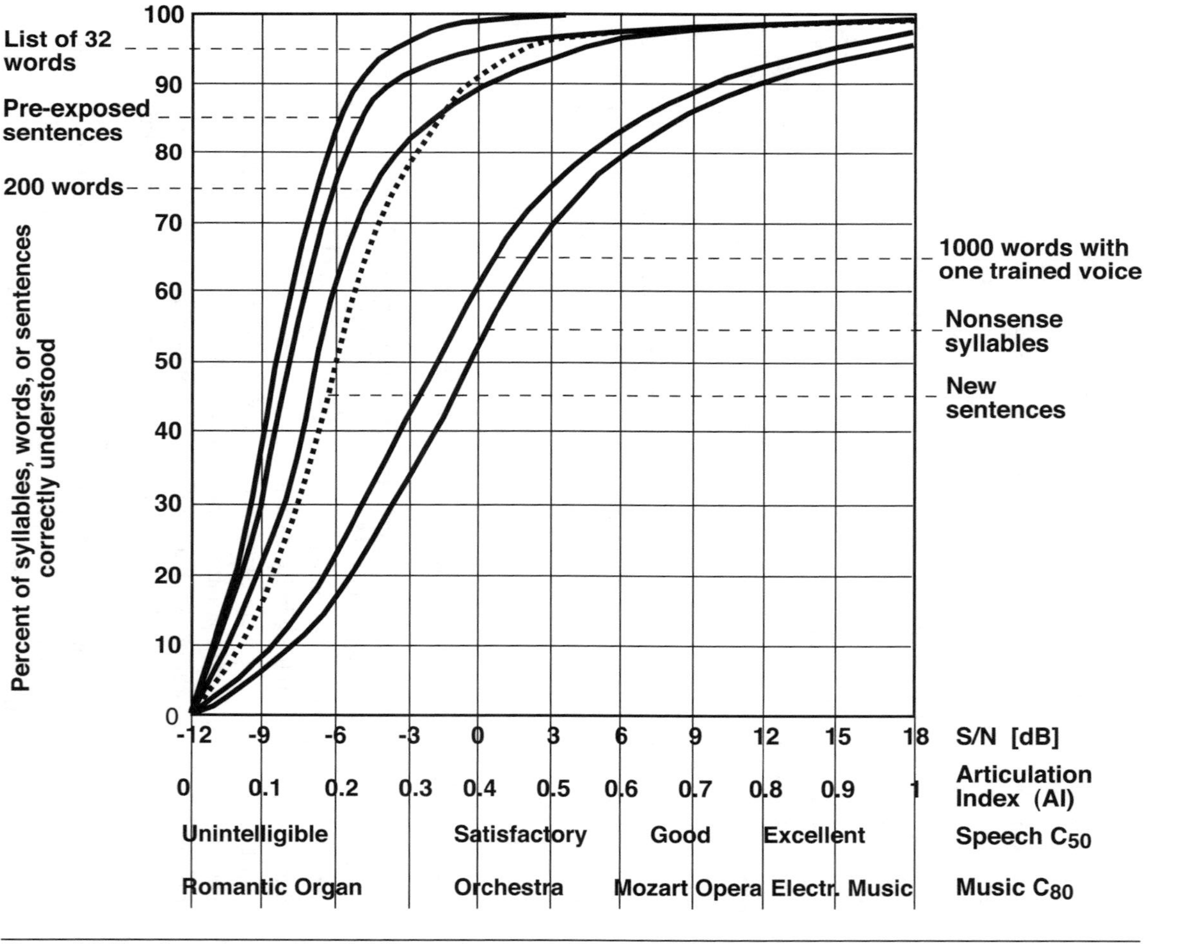

Figure 5.3 The dependence of intelligibility on signal-to-noise ratio, *AI*, C_{50}, and C_{80} (see Reference 5.3).

and word lists but, typically, intelligibility is estimated from knowledge of the signal-to-noise ratio and clarity C_{50} as presented in Figure 5.3.

Speech Transmission Index (*STI*) is based on the modulation characteristics of speech and measures the modulation characteristics of a special signal before transmission and after reception. *STI* can be used to investigate the influence of both speech and reverberation. Because of the extensive subjective testing that accompanied the development of *STI*, it must be regarded as the best metric yet derived for speech intelligibility. Using a computer, the *STI* value can be determined from a measurement of the impulse response of a system and the knowledge of the background noise spectrum.

The relationships between *STI* and various types of speech intelligibility tests are shown in Figure 5.4. Typically, one strives for *STI* values for auditoria to be above 0.65.

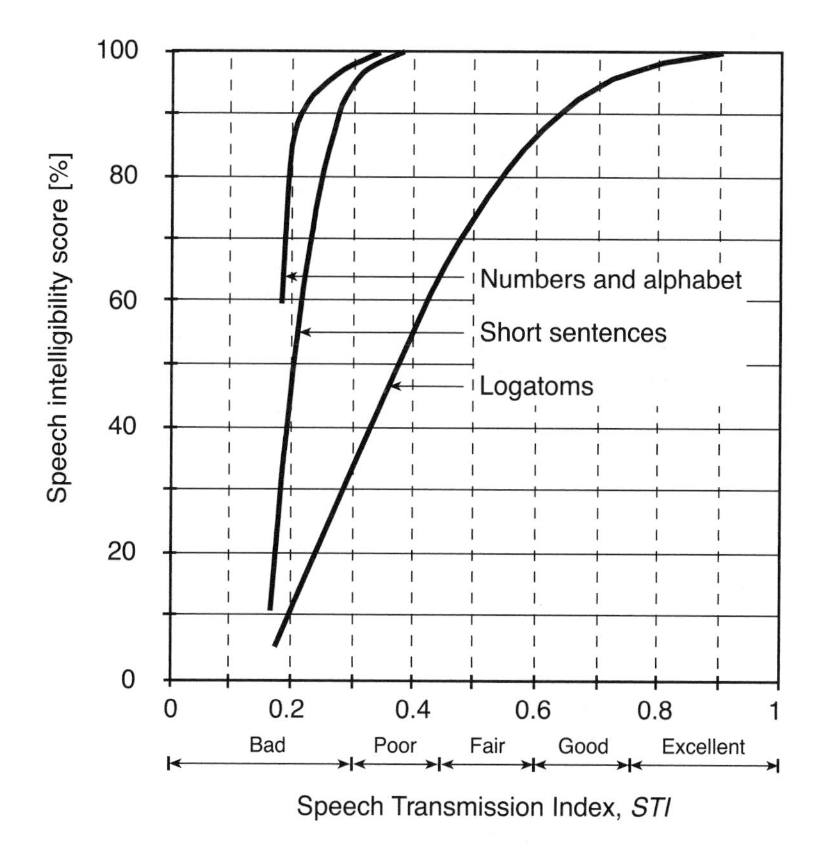

Figure 5.4 Comparison between *STI* results and those of various other speech intelligibility tests (see Reference 5.3).

One problem with *STI* as a metric for the acoustical quality of auditoria is that it does not include the effects of binaural hearing. Comparing listening to natural speech in auditoria to *STI* measurement results, one quickly finds that very different speech quality situations can have approximately the same *STI* values. *STI* also does not measure the sound quality aspects of speech such as distortion and timbre.

The metric Rapid *STI* (*RASTI*) uses a simplified set of *STI* evaluation features. In most applications to room acoustics it gives results similar to those obtained using *STI*. With current high-speed computers, there is no need to use *RASTI* because *STI* evaluation is virtually instantaneous.

5.9 ADDITIONAL ROOM ACOUSTICS METRICS

5.9.1 Strength Index

A metric designed to estimate the relative loudness of sound in a room is the strength index (*G*), sometimes called the *room gain*. The strength index is related to the intensity of sound at a distance of 10 m from the source and is defined as follows:

$$G = 10\log\left(\frac{\int_0^{\infty} h^2(t)dt}{\int_0^{5} h_A^2(t)dt}\right) \quad [\text{dB}]$$

where $h_A(t)$ is the room-impulse response for the direct sound component only, at 10 m distance in anechoic space. Because it is difficult to find large anechoic spaces, the measurement is taken at a convenient distance and then scaled to compensate for the geometric strength difference due to the difference in distance, assuming a level drop of −6dB per distance doubling.

The strength index metric is somewhat related to the degree of *spaciousness* experienced when listening to sound in a room since it depends on the level difference (*H*) between direct sound and reverberant sound as illustrated in Figure 5.5.

5.9.2 Bass Ratio

Warmth is an important requirement in halls for classical music. It is dependent on the relative energy between reverberant sound at low and medium frequencies, respectively. Because the relative energy is, in turn, dependent on the relative sound absorption, it will also depend on the reverberation time. Bass ratio (*BR*) is a way of expressing this relationship:

$$BR = \frac{RT_{125ob} + RT_{250ob}}{RT_{500ob} + RT_{1000ob}}$$

Expected *BR* values are from 1.1 to 1.25 for rooms with $T_{60} > 2.2$ s and from 1.1 to 1.45 for rooms with $T_{60} < 1.8$ s.

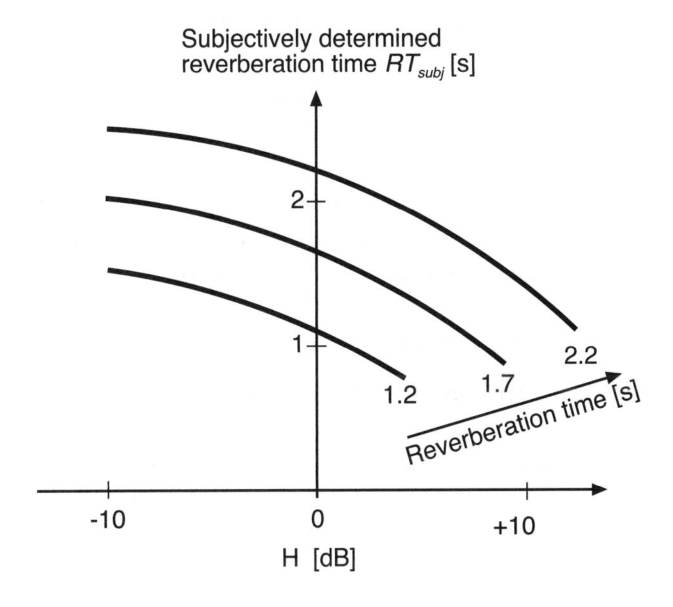

Figure 5.5 Subjectively perceived RT_{subj} as a function of $H = 10 \log(E_{direct}/E_{reverberation})$ with RT as parameter (see Reference 3.2).

5.10 BRILLIANCE

A lack of brilliance can be observed in rooms in which the high-frequency early decay times are much shorter than the mid-frequency early decay times. Two metrics have been suggested for brilliance, $B_{2\,\mathrm{kHz}}$ and $B_{4\,\mathrm{kHz}}$ for the 2 kHz and 4 kHz frequency bands, respectively:

$$B_{2kHz} = \frac{2EDT_{2000ob}}{EDT_{500ob} + EDT_{1000ob}}$$

$$B_{4kHz} = \frac{2EDT_{4000ob}}{EDT_{500ob} + EDT_{1000ob}}$$

Here the EDT indices refer to the octave band frequencies for which they were measured. The typical values for B_{2kob} and B_{4kob} are 0.9 and 0.8 for large rooms, respectively.

5.10.1 Lateral Energy Fraction

It is much easier for hearing to separate between signals if they differ in angle of incidence relative to the median plane. A metric designed to separate between sound arriving in and out of the median plane is the Lateral Energy Fraction (L_f) defined as:

$$L_f = \frac{\int_5^{80} h_L^2(t)dt}{\int_0^{80} h_T^2(t)dt}$$

where $h_L(t)$ is the room-impulse response measured using a bidirectional microphone with its null toward the sound source, and $h_T(t)$ is the room-impulse response measured using an omnidirectional microphone.

5.10.2 Interaural Cross-correlation

The interaural cross-correlation (*IACC*) is an important metric describing the dissimilarity between the left and right ear parts of the binaural room-impulse response. One is usually interested in the reverberation sounding diffuse. Since reverberation diffuseness is closely connected to correlation, it is reasonable to assume that it is advantageous that the late reverberation part of the room-impulse response has as low an interaural cross correlation as possible.

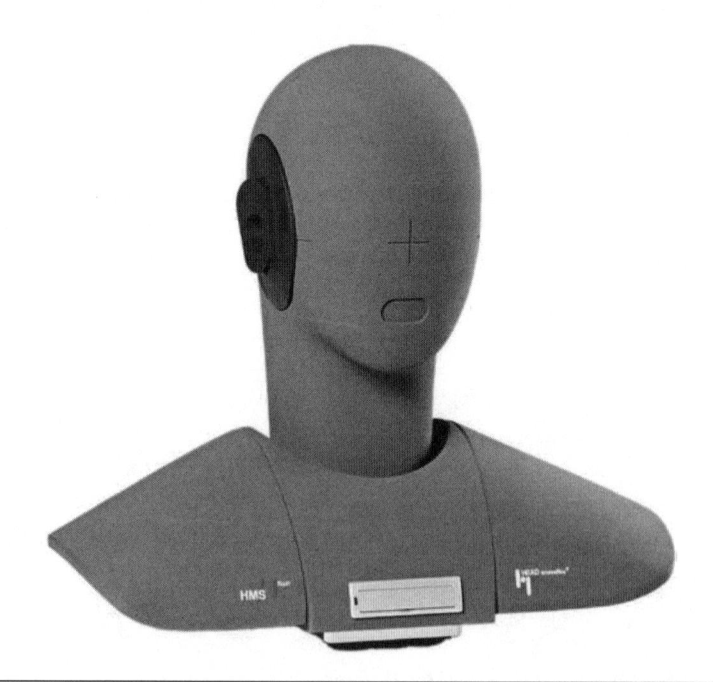

Figure 5.6 A commercial manikin for binaural measurement. (Source: HEAD Acoustics, Germany.)

Originally *IACC* was intended to be used with binaural recording of *running* music so that the metric described what was happening in an actual listening situation.

The interaural cross-correlation is defined as:

$$IACC = \max_{|t|<1ms} \frac{\int_0^{\infty} h_{LE}(t) h_{RE}(t+\tau)\,dt}{\sqrt{\int_0^{\infty} h_{LE}^2(t)\,dt \int_0^{\infty} h_{RE}^2(t)\,dt}}$$

where $h_{LE}(t)$ and $h_{RE}(t)$ are the left ear and right ear measured room-impulse responses, respectively. These are measured using microphones mounted in the ears of a manikin such as the one shown in Figure 5.6.

It is important to note that *IACC* is used with many different types of bandpass-filtered impulse responses. Also, the integration limits vary, and some researchers prefer to analyze *IACC* for impulse response times after 80 ms, for example, i.e., *late IACC*. Care is advised when using *IACC* data as well as when specifying and measuring *IACC*.

6

SIMULATION AND PREDICTION

6.1 SIMULATION AND PREDICTION IN ROOM ACOUSTICS

It is virtually impossible to predict in detail the acoustics of a room by manual means. Even to predict the response of the room in a subjectively plausible manner is difficult. True, one can, with experience, predict the direct sound, important early reflections, and the reverberation time and level, but not subjectively important metrics as the interaural cross correlation.

Acoustical computer-aided design or physical scale modeling is necessary to predict the properties of a room in detail, whether for estimating its speech intelligibility or musical acoustics properties. Figure 6.1 shows a model of a church.

Ultrasonic scale modeling is used similarly, but a physical scale model is built of the room, typically in scales from 1:4 to 1:16, and its acoustical properties are measured. A scale model of the same room as the one in Figure 6.1 is presented in Figure 6.2.

6.2 ULTRASONIC SCALE MODELING

An advantage of physical ultrasonic scale modeling over modeling by computer software is that it is fairly easy to simulate the acoustic conditions even in irregularly shaped rooms or in rooms with unevenly placed sound absorption. In addition, ultrasonic scale modeling allows for inclusion of diffraction and scattering as well as the frequency-dependent sound absorption by various surfaces.

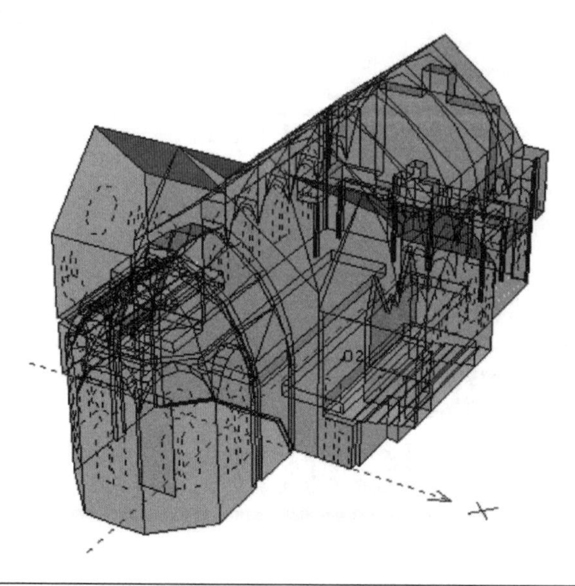

Figure 6.1 A software model of the Örgryte New Church, Gothenburg, Sweden. The model is defined by approximately 500 surfaces.

Figure 6.2 A physical scale model of the Örgryte New Church, Gothenburg, Sweden. The model is built in 1:10 scale and is made up of mostly varnished wood to simulate the hard surfaces of the church.

In ultrasonic scale modeling at 1:m scale, the frequency scale is transposed by a factor of m:1 and the time scale is transposed by a factor of 1:m. After the measurement is completed, the results are converted to full scale and then analyzed.

A major problem in ultrasonic scale modeling is the correct scaling of the attenuation of sound by air. One can approximate it by using dried air or by using a pure gas such as nitrogen in the scale model. Another significant problem can be found in the difficulties in designing sound sources and microphones that can cover the frequency range of interest (typically 200 Hz to 200 kHz in 1:10 scale) with adequate directivity and sensitivity (see Reference 3.1).

6.3 ACOUSTICAL COMPUTER-AIDED DESIGN

Acoustical computer-aided design based on geometrical acoustics is now common in most acoustics consultants' offices. There are many software packages available to the consultant, varying in complexity and features. Some are intended mainly for acoustical purposes, others are focused more on sound reinforcement issues. Using computer-aided design, it is possible to predict a reasonably accurate reverberation time and level, Clarity, Lateral Fraction, Strength Index, and many other room acoustics metrics.

Once the geometrical and acoustical data for the room have been entered into the computational model, one can quickly and easily calculate the way sound from one or several sources arrives at various listening positions.

Most common software for the prediction of room acoustics are based on sound propagation modeling using geometrical acoustics and, therefore, have difficulty in taking diffraction, scattering and real sound-absorbing surfaces into account. Some software will approximate the influence of these acoustic phenomena and thus yield improved modeling.

There are three basic types of mathematical models used in room acoustics modeling-based geometrical acoustics: (1) ray tracing, (2) mirror image prediction, and (3) hybrid methods that combine the first two.

Modeling using the ray tracing method typically simulates a large number of rays (1000 to 1,000,000) sent out from a sound source. The sound source can be assumed to be omnidirectional or directional. The software model then follows each ray on its path and calculates how the ray is reflected by the surfaces of the room and the objects contained in the room. Curved surfaces are generally simulated by piecewise flat surfaces. By studying the number of rays passing through a test surface or test volume—such as a sphere—at the listening position of interest, one can calculate the room-impulse response at that position.

Modeling of sound propagation using the mirror image method assumes that sound is reflected by a flat, hard surface as if coming from a mirror source behind the surface. One needs to calculate the various mirror images created by the flat surfaces of the room and its objects. For a room having a complex geometry, it may be necessary to calculate many millions of mirror sources because sound is reflected multiple times by the various surfaces. The relevant sources must then be sorted out using rules for visibility, etc. A disadvantage of the method is that it does not allow for easy inclusion of scattering and diffraction.

In hybrid models, ray tracing is initially used to find the possible sound paths and then mirror image modeling is applied to find the exact sound paths.

Both ray tracing and mirror image modeling are nearly equal in regard to the computationally intensity. Neither method allows for exact modeling because geometrical acoustics is used. Finding effective modeling methods for absorption, diffraction, and scattering remains a problem. The late part of the room response, corresponding to the late part of the reverberation, is usually too computationally intensive to be handled by a direct approach and, generally, various approximations are used. Some of these may be based on the use of various statistical assumptions (see Reference 6.1).

6.4 AURALIZATION

There are two modes of acoustical computer-aided design. The original one calculates values of various metrics to describe the acoustical quality of rooms as described initially. The second mode is called *auralization* or audible sound-field simulation (see Reference 6.2).

Auralization is a term used in analogy with visualization to describe rendering audible auditorium acoustics and loudspeaker installations. The ambition is to make it possible to listen to the sounds in the environment without being in the actual environment. Auralization provides tools to model a room's acoustical quality adequately in an immediate and obvious way. Typically, the auralization simulations are listened to using headphones, a.k.a. binaural sound reproduction.

Three basic techniques are currently used for auralization. All systems are based on approximations of the properties of the sound source, the room, and the listener:

1. Indirect acoustic scale model auralization uses a physical scale model of the room. The room-impulse properties of this scale model are measured using ultrasonic techniques. Digital signal processing is then used to make the room-impulse responses audible.
2. In fully computed auralization, the sound transmission properties of models of rooms are studied using computer software. This is the method currently most favored because of its flexibility and relatively low cost.

3. A combination of computer prediction, multiple channel convolution, and a multiple-speaker array yields computed multi-loudspeaker auralization.

Auralization was first done using a scale modeling approach. The principle of this approach is illustrated in Figure 6.3. The techniques use a scaled-down human figure—a manikin—with microphones at the ears. The main difficulty is obtaining a suitably scaled manikin with torso, head, and ears, all correctly modeled to scale; additional difficulties lie in modeling seat absorption, etc. A library of scale model absorbers is needed. Once the two sound signals at the manikin's ears have been obtained, the room-impulse responses are processed in the computer with the desired anechoic audio signal and one can listen to what the sound will be in the modeled room.

Computer auralization of an environment will start by entering data for the geometry of the room and the objects in the room into a computer database to be used by a mathematical model based on ray tracing or mirror image modeling. The software will calculate the sound that will be received at the ears. The approach is shown schematically in Figure 6.4.

There are two main presentation methods used for computerized auralization: binaural playback and playback by surround sound systems, including *Ambisonics*. Binaural playback can be accomplished either directly using headphones or by loudspeakers using crosstalk cancellation. For auralization presented binaurally, the software contains information on how sound arriving from any angle is affected by the presence of the head and torso and how it will be received at the ears of the listener. The frequency responses of the sound pressure at the left and right ears relative to the pressure in absence of the body are called the head-related transfer functions (*HRTFs*).

The binaural room response to sound can then be convolved with anechoically recorded speech and music so that one can listen to what the sound of the room would be. The convolution can be accomplished by using software in a general purpose computer or by a dedicated hardware convolver. The convolved binaural signals can then be played back over headphones or further processed by a crosstalk cancellation network to allow for loudspeaker presentation.

In practice, crosstalk cancellation requires a quiet room in which the reverberation time is short. The aim is to provide the listener or listeners with sound signals at the ears corresponding to those that would be obtained when listening directly using headphones. This requires that the leakage of sound between the ears due to the loudspeaker presentation is negated using cancellation signals generated using knowledge about the transfer paths between the loudspeakers and the ears. Presentation using crosstalk cancellation generally gives better results than ordinary headphone presentation, if head movement is held to a minimum.

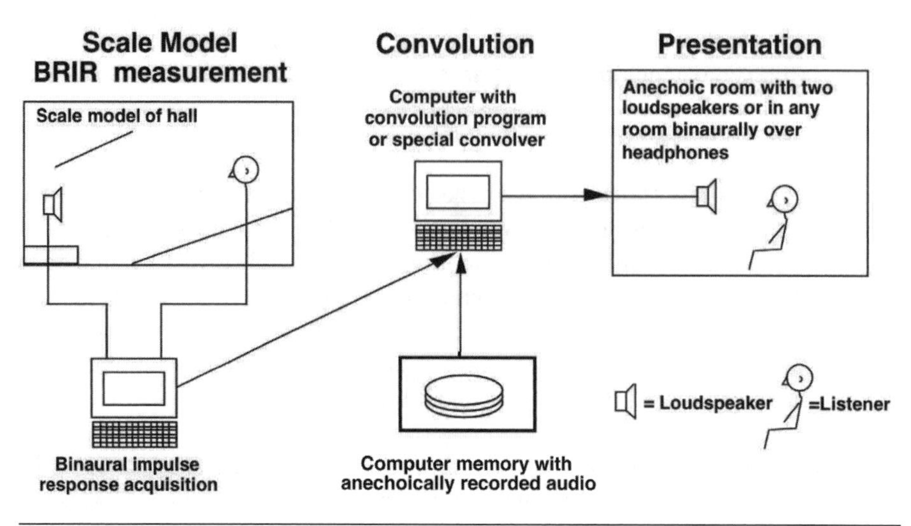

Figure 6.3 Basic principle of indirect acoustic scale model auralization.

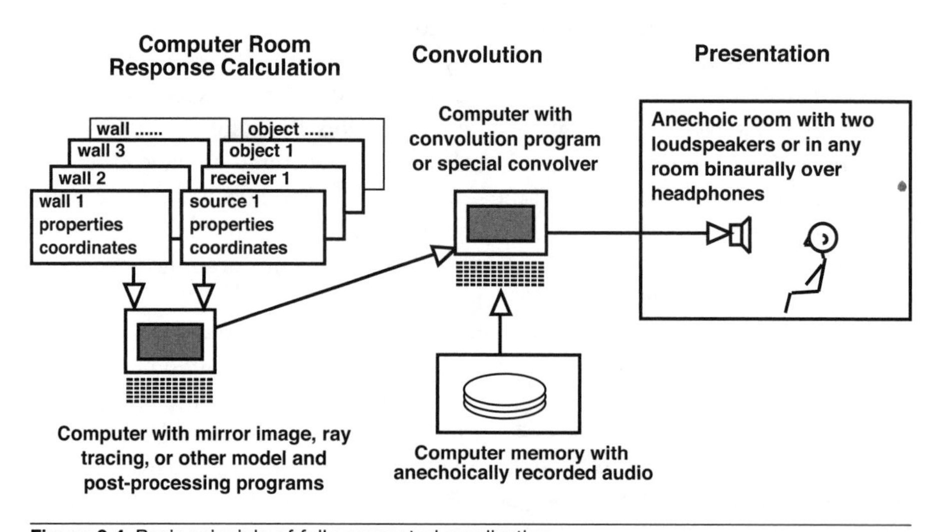

Figure 6.4 Basic principle of fully computed auralization.

Computed multi-loudspeaker auralization is essentially the same as that just described but uses playback over a surround sound system instead of binaurally. These surround sound systems need at least four loudspeakers; 4 to 10 loudspeakers are typically used. The signals are routed to the various loudspeakers according to the desired direction of sound to match the computer-determined direction of the direct sound or the particular reflection. Advanced localization technology is used to determine the signal to two or more loudspeakers for a direction that does not match the exact direction of a single loudspeaker.

The advantages of auralization are many; by listening to auralization of rooms, costly mistakes in planning and building can be avoided and, additionally, various wall treatments and seating arrangements can be investigated. Auralization can also be used to study the properties of sound reinforcement systems to both refine a basic design and to reduce adjustment time at building completion.

7

PLANNING FOR GOOD ROOM ACOUSTICS

7.1 INTRODUCTION

The objective of room acoustics planning and design is to provide rooms that allow good communication for speech and music between speakers, singers, and instruments and the congregants/listeners. In the worship space, congregants participate as sound sources, communicating also with one another by prayer and song, therefore, the send-receive situation is different from that of a theater, concert hall, or similar space.

Room acoustics planning also includes the control of acoustical defects such as noise, echo, and rattle. Noise control is discussed in Chapter 8 and echo and other acoustical problems are discussed later in this chapter.

Background noise levels must be sufficiently low to allow good communication. In rooms intended for worship, low levels of background noise are needed because the contrast between the sound level of the spoken and musical message and the silence during pauses enhances drama. Low background noise also will allow worshippers to hear the full, dynamic range and spatial qualities of the reverberation. The sound quality of background noise will vary, depending on individual perference and on whether the noise is due to heating, ventilation, air conditioning (usually called HVAC noise), to other activities in the building, to traffic, or to neighboring industry. This leads to different noise criteria being necessary, depending on the noise and the use of the room.

For worship spaces, the placement of the worshippers' seating, the choir, and the organ are of special interest. Speakers, singers, instrumentalists, and worshippers may have conflicting requirements regarding suitable acoustics. Individual preference may be substantially different. Solo performers such as speakers and singers will usually want to sense the acoustic response of the auditorium, whereas choirs, musicians playing in groups, or orchestras will put more emphasis on the way the room allows the members to interact. Organ players typically want the room to enhance the size of the instruments. Without reverberation, the sound quality of speech and music would be dependent on the directivity of the sound source and would lose beauty and emotional power as well.

The way sound is distributed in the room is not only expressed by sound level and sound level distributions but also by timbre and temporal characteristics. The temporal characteristics are a result of the room size and shape as well as the way the audience and other sound absorbing and scattering areas are distributed over the room's boundaries. Acoustic planning and design must be introduced in the early sketching stage when developing a new worship space, therefore, cooperation between the architect and acoustician is necessary for effective acoustics. If a pipe organ or large electronic instruments are to be installed, the builder of those instruments should also be involved from the beginning of the design process. The possibilities of changing the general acoustics of a room without building a new interior shell or implementing other drastic changes, such as suspending sound-reflecting panels and installing sound-diffusing or absorbing wall surfaces, are limited once the room exists physically. For rooms that already exist, sound systems that enhance clarity or reverberation, as described in Chapter 10, may be the most cost-effective way to improve acoustics.

The quality of room acoustics is difficult to measure. Chapter 5 discusses some metrics for room acoustics quality. Among these, reverberation time (*RT*) and Clarity are the most important. Room acoustic metrics often require considerable experience and expertise to be applied correctly. Frequently, listening (hearing, itself, is an important measurement device) and subjective judgment by experts is the only way to assess the room acoustics quality.

During the planning of the worship space, it is important that acoustic faults such as noise and echo are eliminated and that the desired balance between direct, envelopmental, and reverberant sound is achieved for clarity and reverberation. This can be accomplished both by passive and by active means. Passive means include: (1) choice of seating area size and arrangement, (2) design of room shape (size, plans, and sections), (3) choice of ceiling height (i.e., room volume), and (4) choice of sound-absorbing and scattering properties of walls, ceiling, floor, and seating area. Active means include sound reinforcement and reverberation enhancement systems. Whereas the influence of all of these factors are interdependent, one can say that room shape and seating primarily determine direct sound and early-reflected sound for the worshippers, and that room volume, absorptive and scattering properties of room surfaces primarily determine the reverberant sound.

7.2 PSYCHOACOUSTICS: THE PRECEDENCE EFFECT AND BINAURAL UNMASKING

Without the signal processing provided by our hearing that creates the so-called precedence effect, we would have difficulties communicating in rooms. The precedence effect makes it possible for us to hear the direction of the location of the sound source from the arrival characteristics of the direct sound. The precedence effect, sometimes referred to as *the law of the first wave front,* usually works quite well in rooms, auditoria, etc., but in some cases, such as when the sound source is obscured or when there are focusing surfaces, the directional impression can shift in an undesirable way.

Experience has shown that a somewhat later arriving sound due to a reflection (for example, delayed by 20 ms corresponding to an additional 6 m of travel) can be up to 10 dB stronger than the direct sound without affecting the apparent direction of the sound source.

The precedence effect can be beneficial in many sound-system applications in which enhanced clarity is desired in worship spaces. By introducing a small delay in the amplified signal, the directional impression of the sound source will remain unchanged, at the chazzan, priest, or imam.

Binaural unmasking makes it possible for us to separate direct from reverberant sound and makes the sound more clear. It is easy to convince oneself of this effect by simply listening with one ear blocked by a finger in a reverberant room.

7.3 SEATING AREA

Note that there is no worshippers' seating in mosques and some Greek Orthodox churches; in this text, the term seating area has been retained for all denominations to describe the area in which congregants assemble to worship.

Most traditional churches and mosques will require the worshippers to be on a horizontal, plane floor. Evangelical churches will often have a sloped seating area. Synagogues will often feature a horizontal main floor seating area and galleries with sloped seating areas, although many synagogues also have sloped seating on the main floor.

Direct sound is the sound that propagates the shortest route from the sound source to the listener without any blockage such as a barrier or a corner. The strength of the direct sound will be higher the closer the worshippers are to the sound-source (see Chapter 1).

In practice, the first 5 ms of sound to reach the listeners are considered to make up the direct sound that will also include early-reflected sound (arriving within 5 ms of the direct sound) from the floor on which a speaker is standing, from nearby objects, and from the seating and worshippers themselves as shown in Figure 7.1. The latter sound, reflected at a shallow angle, is called grazing reflected sound and cannot be separated from the *true* direct sound because of the temporal masking in our hearing. The interaction between the true direct and the grazing sound leads to a timbre change in the direct sound, causing it to be thinner and less full-bodied. A measurement of the acoustical effect on the spectrum of the direct sound is illustrated in Figure 7.2.

Figure 7.1 Grazing sound reflection is due to sound being reflected by seating and worshippers themselves when the sound source is approximately the same height as the listeners' heads. The dashed lines indicate incident sound at grazing angle and the solid line indicates reflected sound. Interference between the two sounds results in the seat-dip effect at seats distanced from the source (see Figure 7.2).

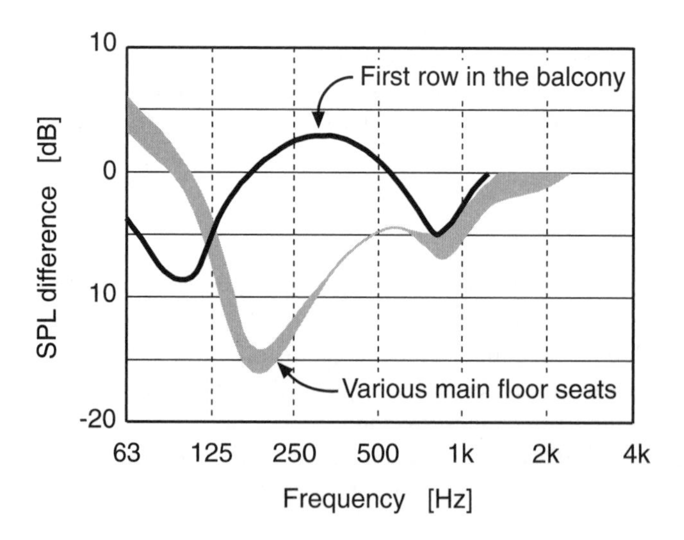

Figure 7.2 Measured early-sound SPLs (in a concert hall) just above audience seating minus calculated SPLs based on spherical divergence of *true* direct sound. Some reinforcement due to grazing-reflected sound gives a positive difference but the seat-dip phenomenon gives a negative difference, adapted from Reference 7.1.

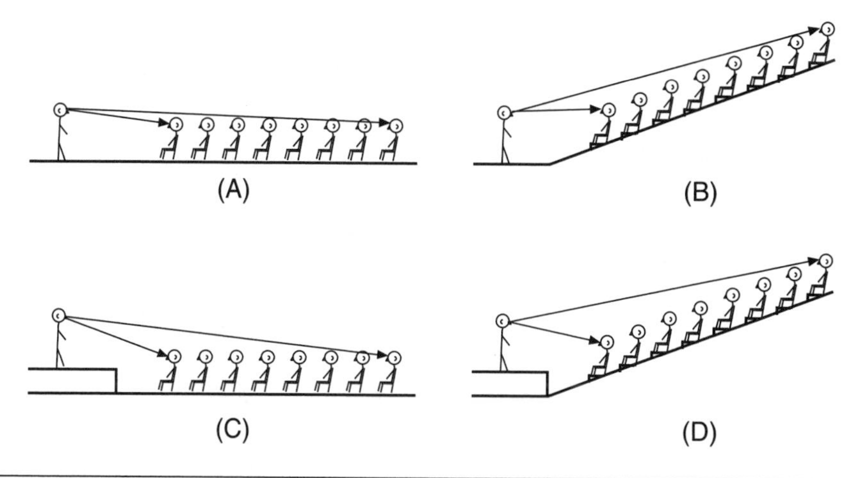

Figure 7.3 Reduction of the seat-dip effect can be achieved by sloping the seating area and by lifting the sound source relative to the plane of worshippers. Case A should be avoided, cases B and C offer improvement. Combination as shown in D offers even greater improvement.

By introducing a sloped seating arrangement, particularly one with an increasing slope, this effect may be minimized because the angle at which sound is reflected by the listeners is less grazing. The effect is also minimized by raising the sound source above the plane of the listeners' heads, as indicated in Figure 7.3, and it is typically absent at the front seats of balconies, which contributes to the excellent sound quality there. For most church organ installations, the effect is also absent because the organ is typically mounted high over the plane of the worshippers as discussed later in this chapter.

7.4 FLOOR PLANS

Is there a preferred floor plan? Also, venues other than worship spaces such as theaters and rooms for music need to be optimized in regard to speech and music sound quality and clarity. Some common theater designs are the proscenium, thrust-stage, and in-the-round designs. Acoustically, worship spaces may be thought of analogously.

An untrained speaker typically has insufficient sound power to reach distances greater than 20 m (60 ft) even in quiet and nonreverberant spaces. One can regard this as the farthest distance from a listener to a speaker in a worship space. Figure 7.4 shows various distributions of the worshippers seating area assuming a maximum distance of 25 m (80 ft) between the center of the wall at the sending end and the farthest point in the seating area.

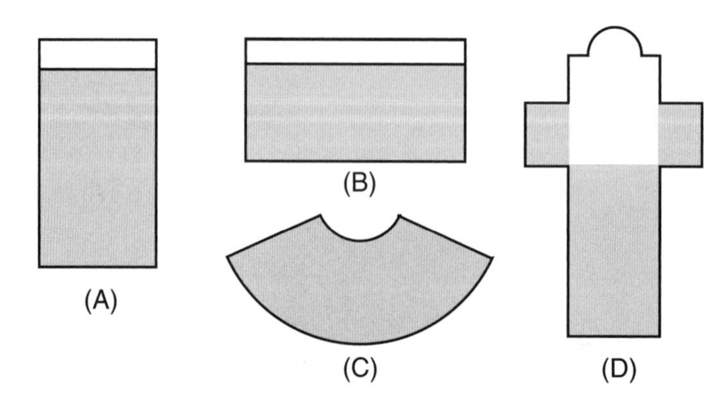

Figure 7.4 Some typical room plans for worship spaces. Grey areas mark out possible seating areas. The basilica (a) and cruciform (d) plans are common for traditional churches. Plan (b), hexagonal and octagonal plans, are often found in mosques. Plan (a) is also a common plan for synagogues, and the fan-shaped plan (c) is a common plan for evangelical churches. (See the chapters on synagogues, churches, and mosques in Part 2 of the book.)

Many traditional churches and cathedrals have a basilica or cruciform floor plan. A cruciform worship space with seating in the transepts also (see Figure 7.5) can be thought of as a variation on the theater thrust-stage, allowing more worshippers to come closer to the speaker, choir, and organ (unless the organ is placed at the back of the nave). For maximum speech intelligibility, each transept may need its own loudspeaker system in addition to the nave's sound system.

The fan-shaped seating plan provides the shortest average distance and the largest seating area within a limited radius from the sound source. However, the directivity of the human voice needs to be taken into account. The voice directivity will lead to less speech consonant power being radiated to the sides. The maximum direction to the sides can be illustrated by the polar diagram shown in Figure 1.10. At ±60° the response at 2 kHz has dropped by approximately 3 dB relative to the front direction. This corresponds to a drop in the distance limit mentioned from approximately 20 m to14 m (65 ft to 45 ft). Because of this directivity, a large fan-shape plan only works with amplification.

Another important effect is that much of the intelligibility of speech is carried by the visual information obtained by watching the speaker's mouth movement, particularly for consonants such as *p*, *b*, and *d*. Lip-reading is much more difficult when watching a speaker from the side rather than from the front.

One must also note that the fan-shaped plan results in echoes from the back wall unless that wall is treated by sound-absorptive or diffusive covering (similar to that in circular, octagonal, and hexagonal spaces discussed below). Both treatments result in shorter reverberation times, more so for absorption (which is

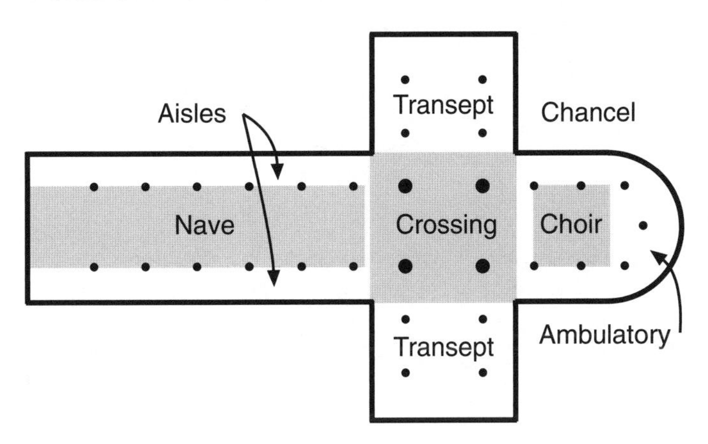

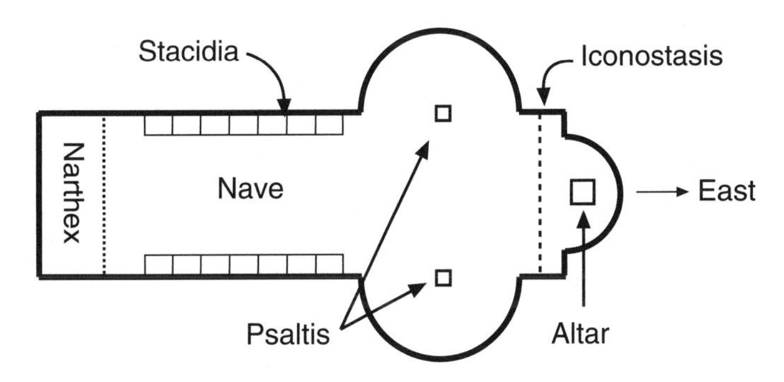

Figure 7.5 Some terms used in describing cruciform churches.

frequently a result because of the medium- or low-ceiling height commonly used with this plan). In the fan-shape seating plan situation, sound reinforcement will be necessary both to overcome the effects of great distance and voice directivity. The same applies to the rhomboid seating plan in which the main sound source is at one of the corners of a square.

In small in-the-round spaces, the average listener will not be more than 7 m (20 ft) away from the speaker, and the direct sound to behind the speaker may be supplemented by useful reflections from the facing wall (see Reference 7.5).

Because of the greater time delays involved, medium-sized and larger spaces having in-the-round plans will suffer unwanted echo and undesired sound concentration because of the wall facing the speaker as shown in Figure 7.6a. An

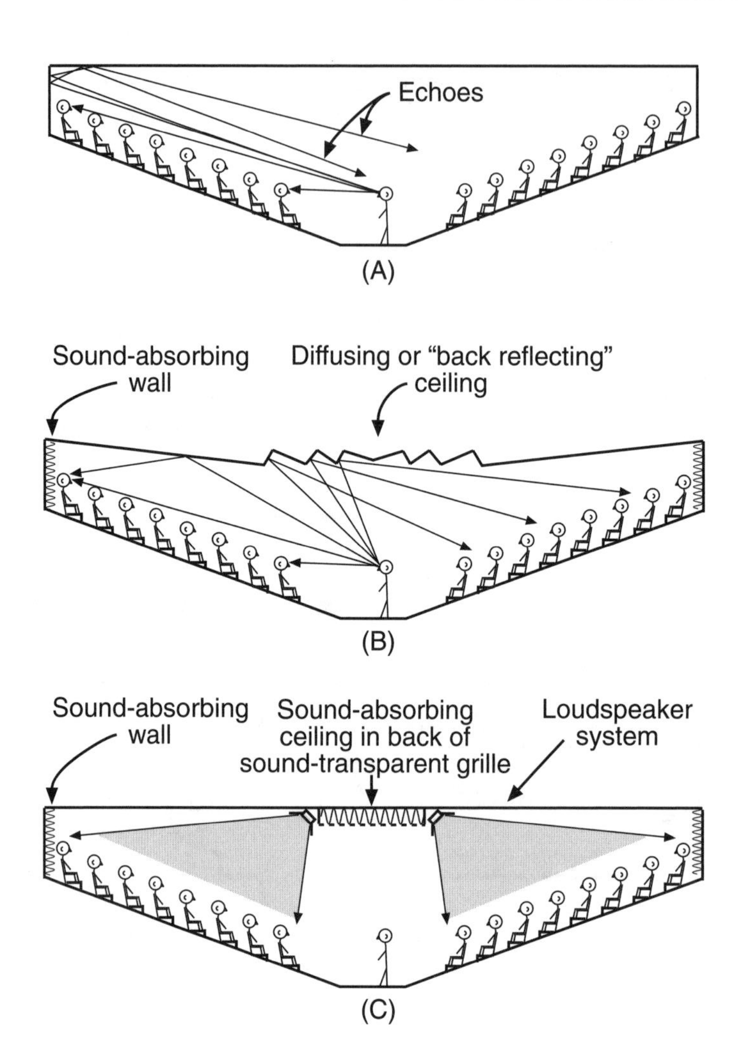

Figure 7.6 (a) Medium-sized and large circular, octagonal, and hexagonal worship spaces using an in-the-round plan frequently suffer echoes as indicated. (b) In medium-sized spaces (less than 600 seats), making the perimeter wall sound absorptive and using a back reflecting ceiling may be sufficient. (c) In large spaces a sound reinforcement system will be needed as well as sound-absorbing material above the speaker's platform.

example of how a medium-sized in-the-round worship space (seating less than 600 persons) may be designed is illustrated in Figure 7.6b. The perimeter walls will need sound-absorbing treatment to reduce echo and reverberation and the ceiling should back reflect to enhance clarity for those listeners not facing the speaker. The ceiling should not be more than 7 m (22 ft) above the center floor or the sound reflections returned to the worshippers behind the speakers back will be delayed excessively (see Reference 7.5).

A large circular space is notoriously difficult unless an expertly designed and sound-absorptive treatment is used on focusing areas such as the perimeter walls. It will need sound reinforcement (see Figure 7.6c). Such a design was used for the 48 m diameter 6000 seat circular Church of God of Prophecy, Cleveland, TN. Six central clusters with directional loudspeaker systems provided coverage for all the seats, and a supplementary ceiling loudspeaker system was used for the 24 m diameter stage.

7.5 LENGTHWISE SECTIONS

Both lengthwise and crosswise sections are of interest because they define the room and, thus, how sound will be reflected by the walls and other surfaces. The voice power radiated to the back and the top of the speaker is lost unless reflecting surfaces, which have been carefully chosen, redirect this power for enhanced clarity and speech intelligibility.

In a rectangular plan room with a plane ceiling, the sound incident on it at the middle and back will be reflected to the rear wall. However, reflected a second time by the rear wall, the sound will be late in arriving and contribute detrimentally toward echo and reverberation. Figure 7.7 shows how an overhead reflector can improve distribution of early-reflected sound and how a sound-diffusive rear wall will help avoid echo. Figure 7.8 illustrates the sound-scattering real wall in St. Peter's Episcopal Church, Bay Shore, Long Island, New York, as an example of this approach to control echo and still retain reverberation.

Reflected sound that is delayed no more than 50 ms—has travelled less than 17 m (50 ft) relative to the direct sound—will be useful in contributing to clarity and intelligibility. The choice of speaker platform, rear wall, and reflector geometry will determine the early-sound reflections. In worship spaces primarily intended for orthodox Jewish, Muslim, and modern Christian worship, high clarity and short reverberation times will be strived for. Here ceiling reflections should be directed down toward the congregants so that the direct and early-reflected to reverberant sound energy ratio is maximized (considering reverberation as a form of noise).

Traditional Christian liturgical music (organ and classical orchestra) requires relatively long reverberation time for optimal listening conditions, whereas speech and modern evangelical music sound better with much shorter reverberation times.

By shaping the ceiling and walls appropriately, it is possible to combine clarity with reverberation so that this is a useful design for large traditional churches and nonorthodox synagogues that should be designed for optimal music acoustics.

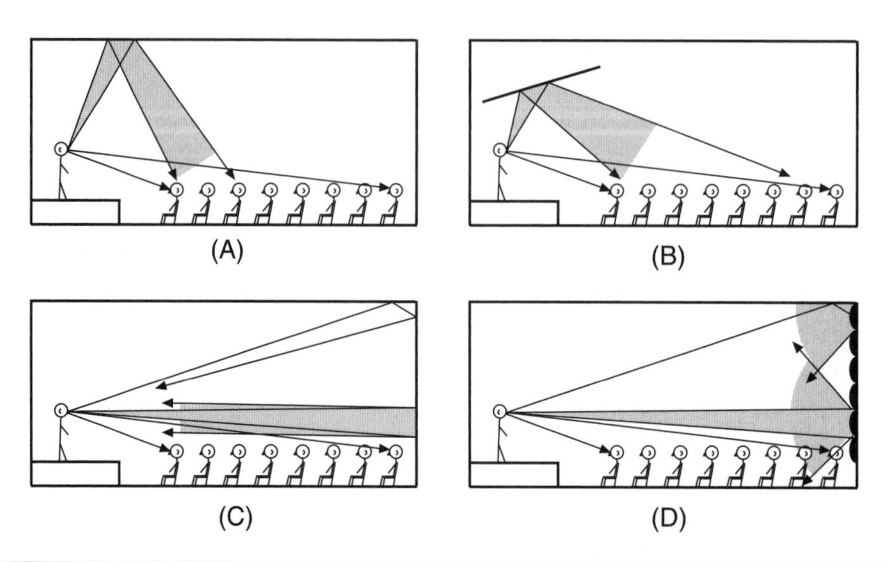

Figure 7.7 (a) and (b) The addition of reflectors can be useful to enhance the clarity at distant seats. (c) In long rooms, a specularly reflecting rear wall, as well as a reflective ceiling/wall corner, may lead to severe echoes in rooms that have a fan-shaped or circular plan. (d) There the rear wall can be made sound absorptive (as shown in Figure 7.6), if a shorter reverberation time is acceptable, or diffusive.

Figure 7.8 The rear wall echo control in St. Peter's Episcopal Church, Bay Shore, LI. Note antiphonal trumpet for Gress-Miles pipe organ, and ship model as the symbol for fishers (photo: David L. Klepper with Klepper Marshall King).

The same approach may be used in orthodox synagogues in which there will be two main sound source locations. Because the five books of Moses are read from a central platform—the bimah—in orthodox synagogues, the back-reflecting ceiling approach is suitable there as well (see Figure 7.9).

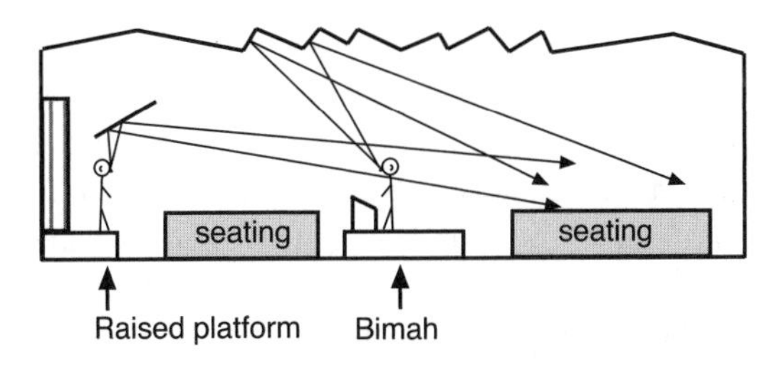

Figure 7.9 Both a sound reflector and a back-reflecting ceiling are useful in orthodox synagogues to improve speech intelligibility at distant seats.

7.6 CROSSWISE SECTIONS

Looking at the crosswise sections, we find that the sound reflected by the ceiling can be directed to reach the listeners directly, to be scattered, or to be reflected to reach the walls and then the listeners. Some alternative ceiling designs are shown in Figure 7.10.

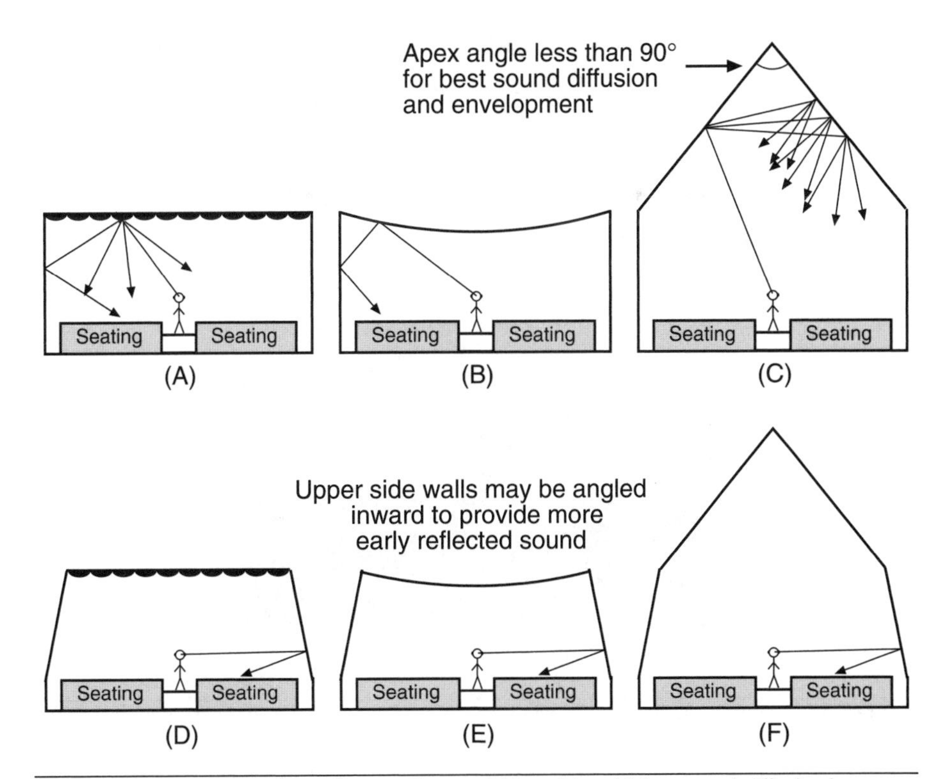

Figure 7.10 Some crosswise ceiling profiles useful to achieve higher sound-diffusion in rooms for music.

In rooms intended for music, the ceiling is sometimes designed to be scattered and to send the sound toward the walls; it is then redirected toward the audience. This makes the sound source appear greater. It also helps blend the sound from the various musical instruments. Usually, this effect is not desired in worship spaces in which the congregants expect the sound to come from the speaker or the singer. In some worship spaces, however, in which there is a choir balcony or gallery and an organ to the back of the congregation, such blending is greatly desired.

By using the curved or diffusive ceiling designs (see Figures 7.10a and 7.10b), incident sound will be redirected to the side walls. Early-reflected sound from the walls will enhance the auditory source width—the perceived width of the sound source.

In some churches and synagogues, the ceiling apex angle is less than 90° (acute angle peak). Sound then will be repeatedly reflected by the ceiling before reaching the worshippers (see Figure 7.10c). Because some sound energy is always scattered at any reflection, this leads to primarily diffused sound reaching the worshippers, resulting in excellent acoustics for music (see Reference 7.6).

Figure 7.11 Interior of Hitchcock Presbyterian Church, Scarsdale, NY, that has a ceiling angle less than 90°, resulting in excellent diffusion and acoustics (photo: David L. Klepper. Sound System Design: Larry S. King. Room Acoustics and Noise Control: L. Gerald Marshall).

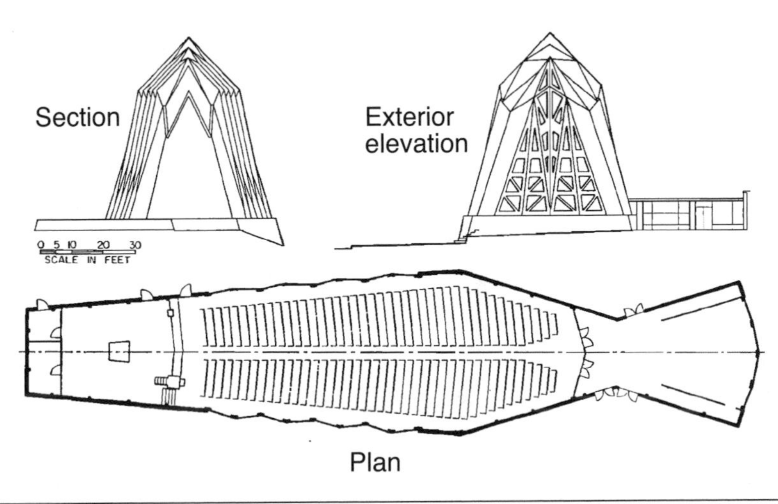

Figure 7.12 The First Presbyterian Church, Stamford, CT, is an example of a worship space combining a small ceiling apex angle with tilted side walls (see Reference 7.6) (Architects: Harrison and Abromovits. Acoustics: Ranger Farrell and Robert Newman at Bolt Beranek and Newman).

An example of such a reverse-V ceiling design is found in the Hitchcock Presbyterian Church, Scarsdale, New York (see Figure 7.11).

If the upper part of the side walls is angled inward more early-reflected sound will reach the listeners. This is one way of controlling the direct and early to reverberant sound ratio. The First Presbyterian Church, Stamford, Connecticut (see Figure 7.12), is an excellent example of this design that combines a small ceiling apex angle with tilted side walls (see Reference 7.7).

If plane ceiling surfaces are made diffusive by decorative elements of sufficient depth such as coffers, cupolas, or arches, the sound will, to a large extent, be scattered and reach the worshippers as early reverberant sound. In many churches, the ceiling is made up of small domes that normally has the center of curvature well above the worshippers and will act as sound scatterers, as discussed later in this chapter.

7.7 PREFERRED REVERBERATION TIME

Experience has shown that the reverberation time suitable for various uses and sizes of rooms can be approximated as illustrated in Figure 7.13.

Generally, one strives to achieve control of the reverberation time to an interval of ± 0.1 s at midrange frequencies. Figure 7.14 shows the average ability to sense a certain change in reverberation time. The common range for reverberation time is 1 to 3 s, in which a reverberation time change of less than 5% can be noticed.

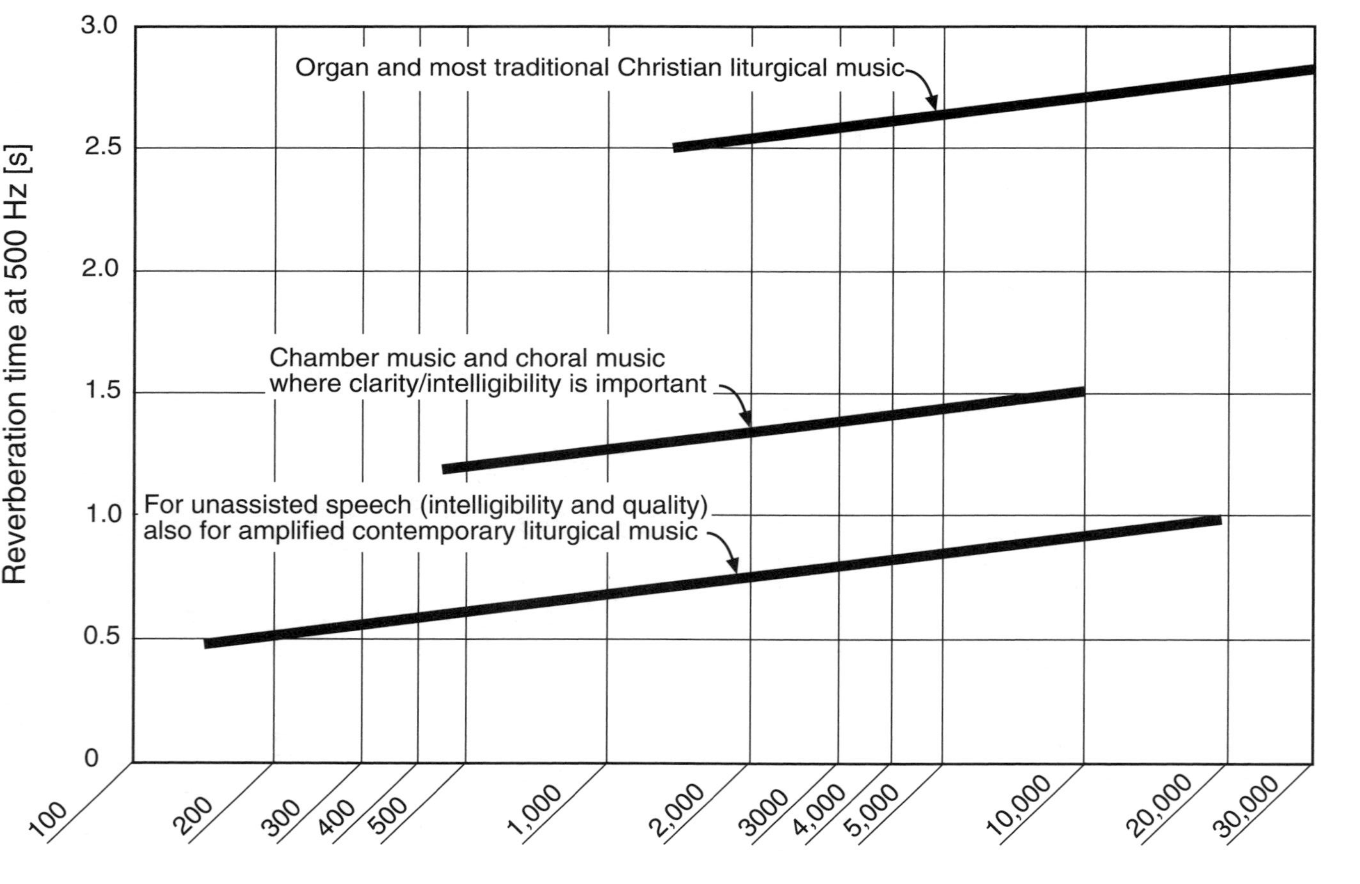

Figure 7.13 Suggested reverberation times for speech and music in worship spaces.

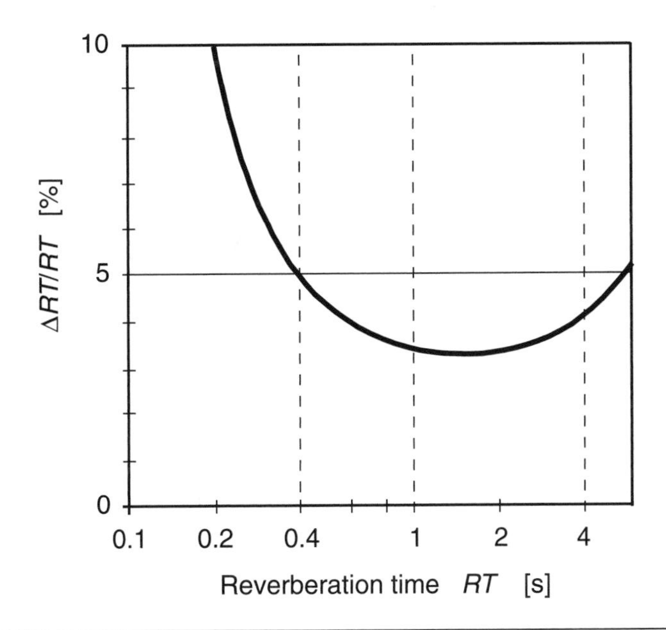

Figure 7.14 Barely noticeable difference in percentage for various reverberation times.

The reverberation time in churches in which organ music is played should increase somewhat at low frequencies from 250 Hz and downward; the reverberation time in the 63 Hz octave band may be approximately 50% higher than in the 1 kHz octave band.

Since the reverberation time requirement for speech is that one should not have any reverberation time increase at low frequencies, the combination of good acoustics for both speech and music may be difficult to achieve unless a reflector or sound system is used for speakers. Excessively sound-absorptive pew cushions or seating will tend to result in *dull* acoustics.

A platform may need to be installed for performances played in the crossing of a church. In mosques, the worshippers are generally sitting on the floor which means that the imam and musicians will be above the worshippers listening plane, improving communication. In large synagogues, the rabbi and chazzan will be on a platform that also helps communication, typically 0.5 to 1 m (1.5 to 3 ft) high.

7.8 COLORATION

Coloration is the term used to describe timbre changes. There are two important types of coloration: one due to the lack of binaural information and one due to the presence of only strong sidewall reflections in the sound field. Two types of reflection patterns leading to coloration are shown in Figure 7.15.

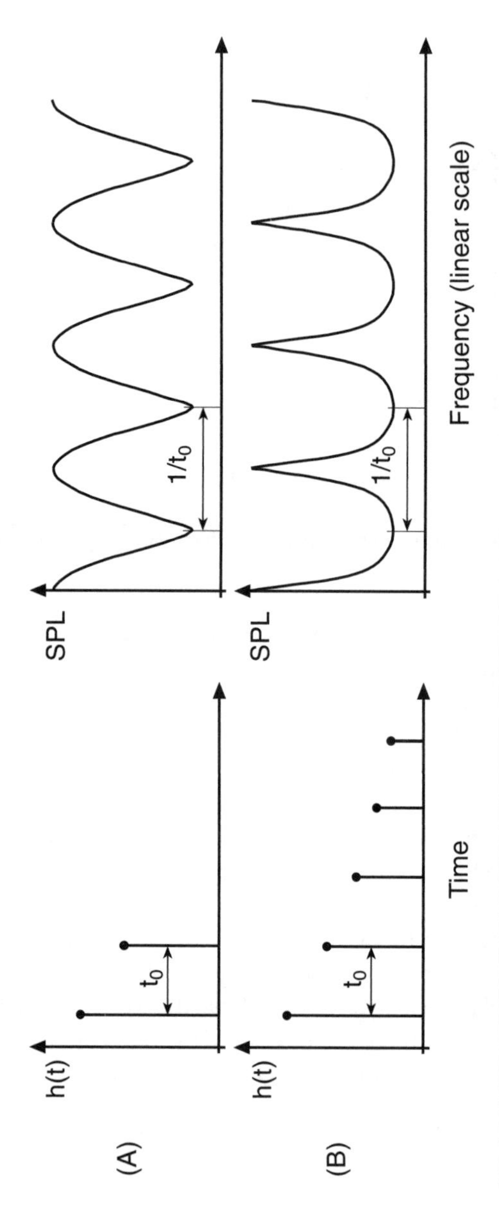

Figure 7.15 (a) Examples of impulse responses h(t) causing coloration and their associated frequency responses. Direct sound and a single strong reflection. (b) Repetitive reflection pattern.

The first kind of coloration (see Figure 7.15a) will generally occur when there is a lack of sound from the sides, for example, when listening to mono sound recordings over headphones. The figure shows how repetitive frequency response variations are introduced by the presence of a strong reflection in the symmetry plane of the head. The precedence effect will be put out of play for a much delayed strong reflection, leading to a shift in the apparent direction of the sound source. For delays longer than 50 ms, echo will occur if the reflected energy is large enough and if there are no reflections filling in between the early sound and the late reflection.

The second, and more serious, type of coloration occurs when one has two highly reflective parallel walls dominating the sound field in a room. This leads to the reflection pattern shown in Figure 7.15b. The acoustic effect is, in this case, called a *flutter echo* or *comb filter effect.* This effect occurs when one has two plane opposing walls, the time delays are small, and there is a lack of other reflected sound. If present, one hears flutter echo easily when clapping one's hands. If the ceiling and wall in a wide room are hard and smooth, one can have the same type of reflection pattern, but in the vertical direction, causing coloration of the first kind.

7.9 ECHO

If the time delay between direct sound and reflection(s) in Figures 7.6a and 7.7a is long enough, and the reflection strong enough, one will hear echo. The sensitivity of the ear to echo is illustrated in Figure 7.16.

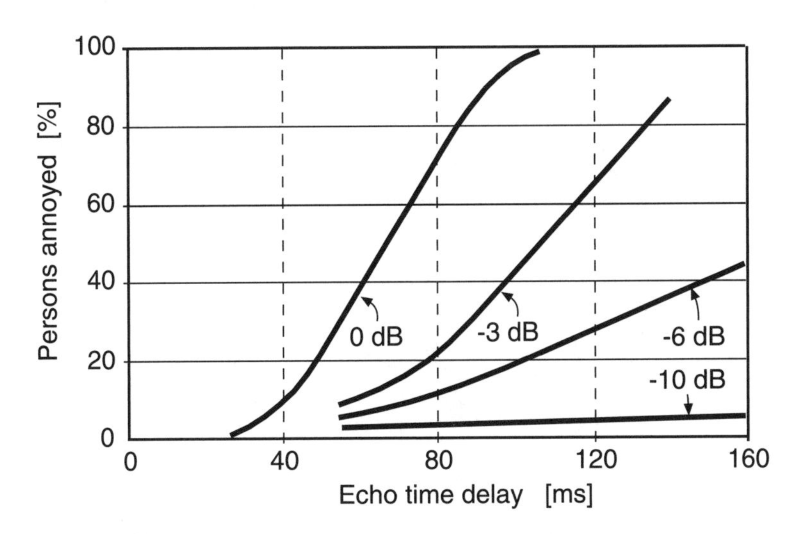

Figure 7.16 Percentage persons annoyed by echo in listening to a direct sound in combination with a delayed sound.

The more rapid the information in the sound, the more annoying the echo is. Slow, low-frequency flue pipe organ tones will not be affected by echo, but speech and fast trumpet passages will. Because of the characteristics of musical instruments and music, echoes are often more annoying at high frequencies than at low frequencies.

7.10 SOME SOUND-REFLECTION PROBLEMS

7.10.1 Domes and Other Curved Surfaces

For echoes to occur, the reflected sound must be fairly strong; this is the case when there are curved surfaces such as domes focusing the reflected sound. Since domes occur frequently in worship space architecture, it is important to know under which conditions echoes occur.

Figure 7.17 shows sound rays from a source at *S* in a room having a circular plan. The numbers identify the points at which rays are reflected. A ray is reflected at the same angle as it is incident against the normal of the circle at that point. After some time, the resulting wave front is the oval shape to the right and focusing has occurred.

Circles, ellipses, and parabolas are the most common curve shapes in architecture and their reflection patterns are illustrated in Figure 7.18. Ellipses have two focal points and parabolas collect an incident plane wave to a focal point.

Domes are frequent architectural elements in worship spaces and can cause multiple reflections against a floor. Two instances of reflection focusing are shown

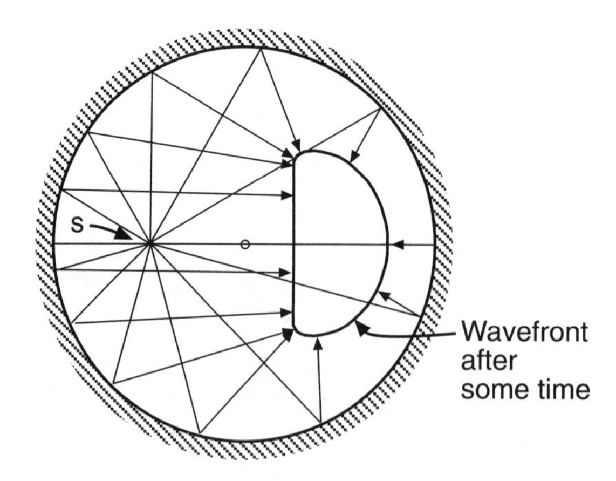

Figure 7.17 Construction of the reflected wave front in a circular plan. All travel paths from the location of the source at *S* are equally long.

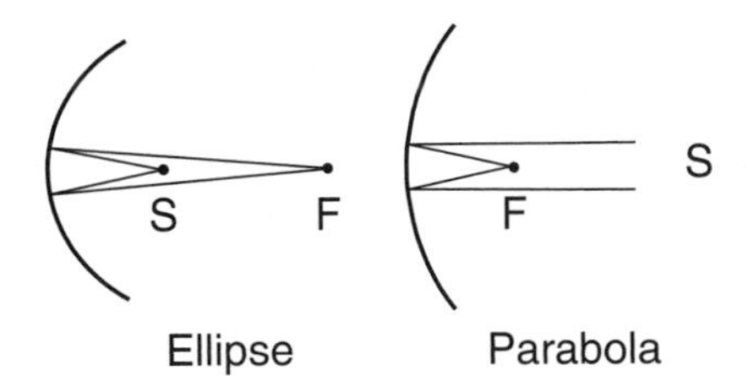

Figure 7.18 Focusing of the reflected wave by ellipses and parabolas (S = source, F = focal point).

in Figure 7.19. In both cases there is a spherical ceiling above a plane floor. The center of curvature is at C, the source at S, and the focal point at F. Domes having their center of curvature below the floor are quite common in orthodox churches. The question then becomes, how can one avoid disturbing focusing. There are several possibilities.

If the worship space is in the planning stage and architecture allows, the center of curvature should be kept well above the heads of the worshippers, because the focusing point will always be far away and not cause harm. The ray construction for this case is presented in Figure 7.20. One should also note that such domes actually scatter sound.

As a rule, to avoid any focusing problem, there should be neither a sound source nor receiver within the full circle formed by the spherical dome. Three

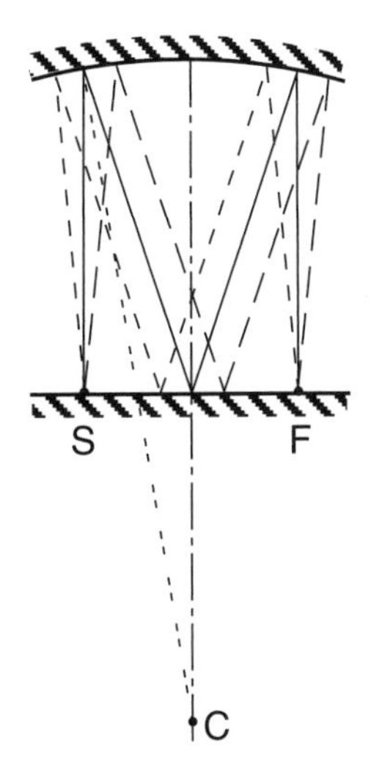

Figure 7.19 Multiple reflections occurring with a spherical shell ceiling because center of curvature is below reflecting floor.

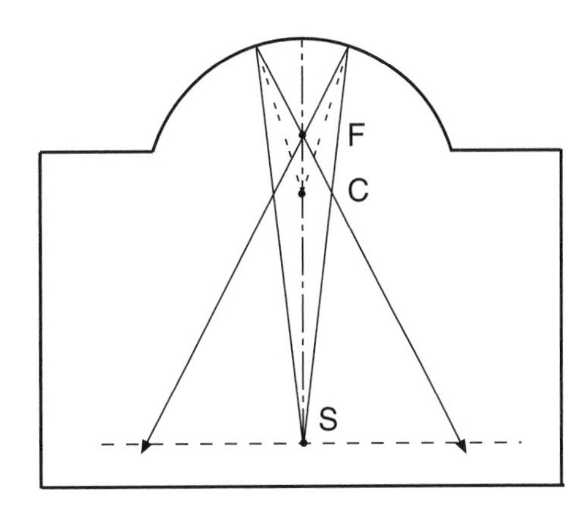

Figure 7.20 In this case, the center of curvature is high because the dome's center of curvature is well above the floor, effectively diffusing sound.

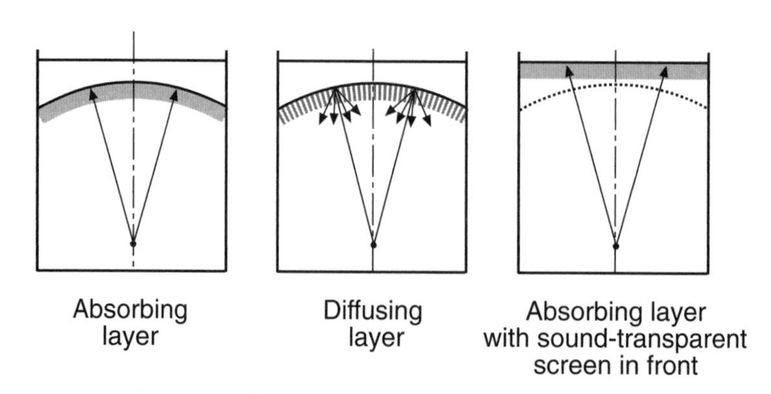

Figure 7.21 Three possibilities for avoiding focusing by a dome.

other possibilities are shown in Figure 7.21: (1) to treat the dome by sound-absorptive material, (2) to make the dome sufficiently diffusive, and (3) to make the dome acoustically transparent, for example, by making it out of perforated metal or other material.

The American church architect, Harold Wagoner, frequently employed a large radius that was three times the height of the laterally curved ceiling and transitioning directly to a small radius that was less than a quarter of the ceiling height at each side giving the appearance of a barrel-vault without its acoustical problems.

7.10.2 Whispering Galleries

The whispering gallery effect is due to high-frequency sound such as a whisper carried around by a curved surface with minimal sound absorption (see Figure 7.22).

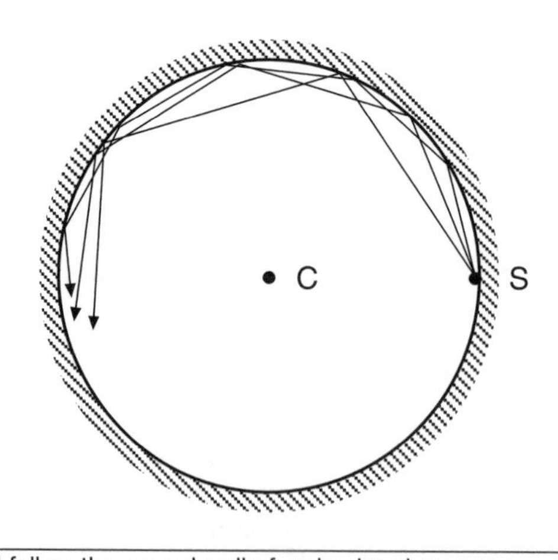

Figure 7.22 Rays will follow the curved wall of a circular plan space.

A classic example of a whispering gallery is found in the Statuary Hall in the United States Capitol, Washington, D.C., but the effect will be found in any room with hard, concave curved walls. It is an acoustical oddity, but it is seldom a problem because it requires both a sound source and a receiver close to the wall.

7.10.3 Pillars

Churches and mosques frequently have large arrays of pillars (see Figures C.9 or M.10). Whereas slim, single pillars will not affect sound appreciably as long as their cross section is small compared to wavelength, there are occasions when pillars cause unwanted acoustical blockage along with their sight blockage. When pillars have cross-section dimensions approximately 0.3 m (1 ft), that is, approximately one wavelength at 1 kHz, or larger, they start to cast a shadow and to reflect sound at frequencies important to speech. Repetitive reflections from a row of pillars may cause an effect similar to that generated by other repetitive sound reflections as discussed previously. Additionally, the reflections reduce speech intelligibility for the worshippers seated in the shadow zones unless suitable reflecting surfaces and/or a good sound system are provided.

At some angle the area behind the pillars will become invisible to the sound source. This is an indication that the direct acoustic path will also be absent for consonants which, of course, results in poor speech intelligibility.

The situation is not as bad for organ music as it is for speech because a small drop in clarity for organ music is not as serious as it is for speech intelligibility. Organ music relies heavily on the diffuse reflections of the room and the reverberation for its acoustic effect. Pipe organs, which are physically large instruments, are often placed at the far end or at the front of a church which means that there will always be numerous reflections both within the organ case and from the side walls surrounding the organ, as well as from the ceiling above the organ (see Section 7.18.2). The pillars virtually will have no influence on the reverberant sound but may still reduce clarity. With the proper design of nearby surfaces, the same effect can be provided for the choir.

7.11 ANNEXES AND DUAL-SLOPE REVERBERATION CURVES

Many churches have a cruciform floor shape which means that there are acoustically four different rooms that are coupled through the crossing (see Figure 7.5). In many churches and cathedrals there are also side chapels (see Figure C.2). This means that there will be many seats for which there are no sight lines, that is, there will be a lack of direct sound and associated early reflections. The rooms will have various early decay times because of the dissimilar geometry and sound difference. If the side rooms—transepts—are mainly used for overflow seating on holidays, then the problem is minor. Otherwise, if the transepts are frequently used, sound systems may need to be utilized to obtain the desired speech intelligibility. Again, it is important to stress that the voice is directional and, thus, requires appropriate compensation.

Because of the differences in reverberation times, there may be dual slope reverberation slopes. The seriousness of this condition depends on subjective preference. From the viewpoint of combining clarity with long reverberation time, the dual slope reverberation may, in fact, be an advantage.

7.12 BALCONIES

Common elements in worship spaces are balconies and galleries. (In this text we will use the terms balcony and gallery interchangeably although architects reserve the term gallery for a worship space balcony that is supported by pillars.) There are many reasons for having balconies; in churches they may be used for added seating capacity and in synagogues and mosques they may provide separate seating for women and men. An extreme case is the mosque shown in Figure M.9 in which men and women worship on two separate floors coupled by way of an open space at the front of the mosque.

When there is a balcony situation, one must consider the sound quality both on the balcony and underneath it. Typically, the ceiling height under the balcony is 2 to 3 m which results in a *home listening* situation, that is, sound similar to that received from a stereo system and coming mainly from the front. The reverberation will also come from the front, as when listening to a stereo system in the home. This means that the desired spatial effect of the reverberation will be absent and, in particular, choir and organ music will suffer. For concert halls it is recommended that the balcony depth should be smaller than the opening height; this is also a good plan for worship spaces in which one strives for natural reverberant sound. For speech intelligibility and clarity, direct and early-reflected sound also should reach those listeners at the far end of the balcony, that is, minimally listeners on the balcony should have an unobstructed view of the speaker or musician.

Four different on top or underneath balcony situations are shown in Figure 7.23. The situation in both 7.23a and b is clearly unacceptable; minimal sound will come in because of the low ceiling height and incoming sound will be attenuated on its way to the far-back listeners. Interaction with congregants, except in close proximity, is also impaired. Because of the increased height of the opening on the balcony in 7.23c, the listeners on the balcony will have improved reverberant sound and early reflections. In 7.23d, the ratio between balcony depth and opening height, both on top and under the balcony, is about unity and will give good acoustical results and will provide visual intimacy as well.

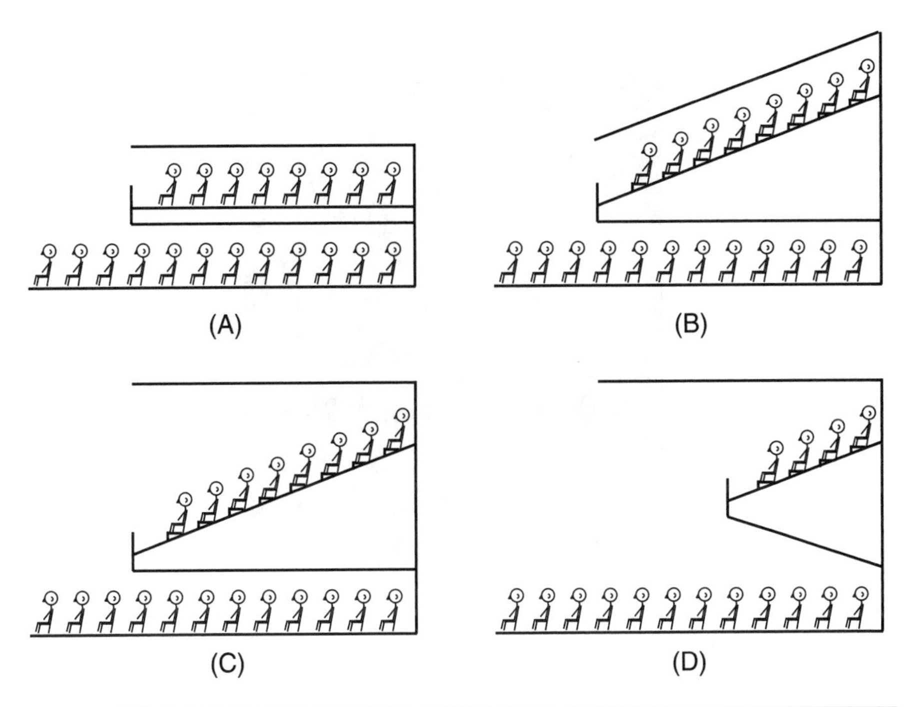

Figure 7.23 Four different balcony situations. Cases (a) and (b) must be avoided, Case (c) offers improvement and Case (d) will give good results both for voice and music.

The photo in Figure 7.24 shows an unusual situation in which a synagogue features two gallery levels. The upper balcony is virtually never used because of the poor speech intelligibility there, a result of the lack of both direct and early-reflected sound.

By having sloping balcony undersides, as shown in Figure 7.23d and Figure S.7a, direct sound, early reflections, and reverberant sound will still reach the congregants. Balcony undersides should be made sound reflective.

Another possibility to eliminate a poor, under-balcony situation is to fix an array of loudspeakers to the ceiling under the balcony. These loudspeakers can then be fed with signals that have been appropriately delayed and reverberated technically, using an electronic reverberation system, or by feeding sound picked up by microphones in more reverberant parts of the church. Obviously, the expense may not be justified if such seating is rarely used, but planning in advance, including conduit and loudspeaker back-boxes for such an installation, might save a great deal of money when the desire to implement such an installation arrives.

Balcony fronts are usually reflective and may need to be angled to avoid echo, or to diffuse incident sound as shown in Figure 7.28b.

Figure 7.24 Synagogue in German 1850s style with two levels of balconies in Gothenburg, Sweden (photo: Mendel Kleiner).

Figure 7.25 Small pulpit canopy reflector in the 17th century Morlanda church, Sweden (photo: Mendel Kleiner).

7.13 REFLECTORS

The beneficial effect of reflectors was discussed earlier. Reflectors are usually used to distribute the sound from a speaker or an orchestra to the audience, or to improve the acoustical conditions for musicians on stage. Reflectors can be planar or curved. Sometimes curved reflectors are used as diffusers.

Figure 7.25 shows a pulpit canopy reflector in a church. The acoustic advantages of such a reflector are small and, with the availability of modern sound systems, there is really no acoustical reason to use such a reflector because it is too small to be acoustically effective. The main positive effect of this canopy will be the added support given to the voice and felt by the priest.

In traditional, orthodox synagogues, the chazzan's voice is directed toward the Aron Ha-Kodesh—in principle, away from most of the worshippers—as indicated in Figure 7.9. In this case, a reflector may be useful; an effective reflector can be made of laminated glass. Figure S.14 shows how the rabbi's pulpit has been used to provide a reflector for the cantor's voice in the synagogue shown in Figure 7.24.

For a reflector to be acoustically reflective, its surface must have dimensions much larger than the wavelength of sound. It must also cover a significant part of the solid angle as seen from the sound source. Additionally, it must have sufficient mass to prevent it from vibrating. Often the criterion is set that the mass per unit area should be at least 20 kg/m^2 (ca. 4 lb/ft^2). To give an appreciable reflection of medium- and high-frequency voice sounds, the mass per unit area can be as low as 2 kg/m^2 and the dimensions 1 m by 1 m (3 ft by 3 ft).

Note that the angle of incidence affects the efficiency of the reflector and that groups of reflectors will show different behaviors from single reflectors. In addition, the reflection properties of reflectors will depend on whatever surface roughness they may have. The acrylic plastic ceiling reflectors shown at the top of the photo in Figure 7.26 are designed to be somewhat diffusive, as well as blocking a focused echo from the domed ceiling above.

Figure 7.26 The translucent plastic ceiling reflectors in the Rodef Shalom Synagogue, Pittsburgh, PA, are designed to be somewhat diffusive and eliminate a focused echo from the dome above (photo: David L. Klepper, Klepper Marshall King, consultants).

Because reflectors are usually relatively small, it is easy to do a provisional installation to convince oneself of the added benefit. Finally, it should be noted that technically the reflection properties of reflectors are best investigated using physical scale modeling.

7.14 BARRIERS AND MECHITZOT

In many cases it is necessary to subdivide a room into visually and acoustically separate volumes. A partial wall, or a barrier, can then be used to achieve a sound shadow. The effectiveness of a barrier is determined by its acoustic and mechanical properties plus the wavelength and the angle of incidence of sound.

For sound in the speech range, that is, 250 Hz to 4 kHz, the wavelengths are between 1.7 m to 0.08 m (6 ft to 3 in). If the barrier height is of the order of 2 m (7 ft), the result is poor shadow action at the lowest-voice frequencies because of diffraction of sound around the barrier.

A thin reflecting barrier will usually have more insertion loss than a thin absorptive barrier. A thick barrier usually results in better insertion loss than a thin barrier because the sound has to propagate over two edges.

It is important that the barrier has sufficient sound insulation so that sound is not transmitted through the barrier. This typically requires a mass per unit area more than 10 kg/m^2 (2 lb/ft^2) for barriers of the type common in open plan offices.

Figure 7.27 A mechitzah is a divider separating men's and women's seating in an orthodox synagogue (photo: Eric M. Saperstein, Artisans of the Valley, Pennington, NJ).

A special type of barrier is the acoustically transparent *mechitza* that is used in mosques and orthodox synagogues to separate men from women in worship. In traditional orthodox synagogues, the mechitza (pl. mechitzot) needs to provide visual isolation and is normally around 2 m high. Figure 7.27 shows a mechitza in an orthodox synagogue. In modern orthodox synagogues, a mechitza need not be a visual barrier and can be as low as 1 m. Only a few conservative synagogues and no reform synagogues use a mechitza. More examples of mechitzot can be found in Part II in the chapter on synagogues.

In the mosque the separation of men and women worshippers is usually achieved by having separate mosques, although, as previously discussed in connection with Figure M.9, some sects only separate men and women so they are out of sight of one another during worship. Liberal Muslims have fully integrated worship.

7.15 DIFFUSERS

A common desire for good room acoustics is to diffuse the sound reflections made by plane surfaces. In most rooms for music and speech, a suitable amount of diffuse reflection will prevent flutter echo and enhance the *smoothness* of sound. For a diffuser to work properly, it needs surface irregularities of approximately a quarter wavelength or larger, but even small surface unevenness is advantageous. Niches about 0.3 to 0.4 m deep (1 to 1.5 ft) are usually sufficient for good results. The irregularities should be randomized for optimum effect, but this is often not possible for architectural reasons. Figure 7.28 shows several types of irregular surfaces that may be used to scatter sound. Nearly any extended roughness deeper than 10 cm (3 in) will add beneficial diffusion.

An acoustic effect of diffusion is that the diffusive surface seems to *disappear*, but that the room still sounds reverberant. Diffusers are common building elements in recording and monitoring studios. Before installing a quadratic residue diffuser (QRD), or other mathematically defined diffusers, one should discuss the possibility of designing custom with the architect—architecturally pleasing unevenness on walls and ceiling.

7.16 TEMPORARY STRUCTURES AND TENTS

Occasionally it will be necessary to set up a temporary structure to be used as a worship space. This may be when a venue not intended for worship is to be used or when one wants to conduct religious worship out of doors. Zoning regulations may also prevent a conventional building. In dry climates, one can make do with

(a)

(b)

(c)

Figure 7.28 (a) Use of polycylindrical diffusors in the organ recital hall at the Gothenburg School of Music, Sweden (Photo: Mendel Kleiner). (b) Use of QRD diffusors on the balcony front of the Clear Lake United Methodist Church, Clear Lake, TX (Photo: David L. Klepper, Klepper Marshall King, consultants). (c) Use of rococo interior of the basilica in the Ottobeuren Benedictine abbey, located in Memmingen, Germany (photo: Johan Norrback, GOArt).

only a sound system; in places with rain, it may be necessary to add a tent or similar building. Figure 7.29 shows the interior of the Benedict Music Tent, Aspen, Colorado, which is a permanent outdoor structure used for summer concerts.

Tents and marquees offer both advantages and disadvantages from the viewpoint of acoustics. Because of the low mass per unit area of the canvas, the *wall* will have poor sound isolation, particularly at low frequencies. This may be of concern to the surroundings because if amplified speech and music are used in the tent, the sound levels can become high outside of the tent.

Typically, the canvas will be made of plastic or some water repellent synthetic cloth. Natural canvas will start to rot and leak after a few years, and modern tent structures are frequently made of Teflon-coated fiberglass. The structure shown in Figure 7.29 uses a canvas that has a surface mass of approximately 2 kg/m^2 (ca. 6 oz/ft^2). The acoustic behavior of the fabric is determined by mass, tension, and damping. The mass will vary with precipitation. A wet fabric has higher mass per unit area. Rain on the fabric will cause considerable noise radiation from the canvas. Normally, the resonances of the fabric pieces are of minor acoustical interest because they are well damped by the air.

Looking at the room acoustics of the tent, one notices that most tents have surfaces that are convex (from the inside). This means that they are diffusive so early reflections will tend to be nondistinct and the reverberation sound will be well mixed; however, the acoustical properties of the floor or ground may be rather absorptive at medium and high frequencies. It is only in the case of the concrete or paved floor that it will be fully reflective.

Figure 7.29 The Benedict Music Tent, Aspen, CO, uses Teflon-coated fiberglass fabric. In addition, conventional reflectors are used to provide suitable early reflections (photo: Kirkegaard Associates, Chicago, IL).

Using the curves in Figure 4.3 one finds that, for a canvas having a mass per unit area of 2 kg/m^2 (6 oz/ft^2), the sound-absorption coefficient is less than 0.2 only over 200 Hz. This means that amplification and loudspeaker systems would be needed to provide low-frequency sound. A mass per unit area of 10 kg/m^2 (2 lb/ft^2) would essentially eliminate the problem and provide sufficiently full range sound reflection for most purposes.

7.17 ROOMS FOR SPEECH

Good speech intelligibility is not the same as naturally sounding speech. One can increase the intelligibility of amplified speech, particularly in noisy surroundings, by various types of signal processing such as spectral shaping and compression.

As discussed in Chapter 2, we can regard speech as a modulated signal (see Figure 2.5) with a complex tonal or noisy spectrum primarily covering the frequency range 0.25 to 4 kHz that is modulated by frequencies in the 0.2 to 8 Hz range. Reverberation and noise will decrease the modulation. A reduction in modulation results in a reduction of speech intelligibility. For speech modulation that does not affect the listener by reverberation, the reverberation time must be short and the ratio between direct and early-reflected sound to late-reverberant sound must be high.

There is, however, a practical limit to shortening the reverberation time. It is expensive to add sound absorbers to a room and it may be architecturally unacceptable. In addition, rooms are expected to have a certain reverberation time and level, determined by tradition and visual aspects (room purpose, shape, and volume, for example).

Speech intelligibility can be shown to increase with increasing reverberation time for short reverberation times in the range of up to 0.5 s. The reason is that sound absorption is coupled to both reverberation time and sound level. In addition, early-sound reflections by more reflective walls help to eliminate the influence of speaker sound radiation directivity on the sound pressure. The output of the human voice varies approximately 6 to 10 dB with angle in the frequency range most important for speech intelligibility.

These are some good rules for auditoria used primarily for the spoken word:

- Keep the travel paths of direct sound and important early reflections short. It is difficult for an untrained speaker to reach audiences with sufficient speech intelligibility at distances over 20 m (60 ft) without the use of amplification.
- There must be enough early reflections to provide sufficient sound at the listeners while at the same time not feeding sound energy into

the late part of the reverberant sound field. It is advantageous to strive for early-sound reflections to arrive close to the horizontal plane, rather than from above, to have good speech sound quality. However, sometimes it is necessary to make use of overhead sound reflectors as shown in Figure 7.24 or 7.25.

- Mirror-like (specular) reflections must not exceed the level of the reverberant sound if they arrive more than 30 ms after the arrival of the direct sound; such reflections may be perceived as echoes. As discussed previously, focusing by concave surfaces must be avoided.
- The human voice is directional. In small rooms, including classrooms and small auditoria having relatively hard walls, the speaker usually will be quite close to some sound-reflecting surfaces. This leads to strong early reflections that will add to the direct sound in such a way as to increase the speech intelligibility. In large auditoria such as worship spaces, assembly, and lecture halls, it is important to place the seating area within approximately a 120° arc from the speaker. The effective speech intelligibility will increase when the listeners can see the talker's lip movements.
- For best speech quality, the reverberation time should be in the range of 0.6 to 1.0 seconds, increasing with the size of the room (see Figure 7.13). It is possible to combine this with the desire to have long reverberation time for music by using sound systems or designing the room in such a way that the reverberation curve has dual slopes. For speech the reverberation time should also be constant to within 10% over the 125 Hz to 8 kHz octave bands for small- and medium-sized rooms. The necessary room volume depends on the sound absorption by the audience. Typically, each person requires approximately 0.75 to 1 m^2 (7 to 9 ft^2) of floor space. This results in a ceiling height in the range of 3 m (10 ft). Sound absorbers should be placed so that they do not interfere with the propagation of important early-sound reflections. This means that one should leave the central part of the ceiling nonabsorptive in auditoria, classrooms, etc.
- For small studios and control rooms (such as those used at radio stations having volumes smaller than 25 m^3 (750 ft^3), it is common to reduce the reverberation time to the range 0.2 to 0.4 s for a neutral sound. It is sometimes also necessary to control the resonant modes at low frequencies using various types of tuned resonant sound absorbers, *bass traps*.

7.18 ROOMS FOR MUSIC

7.18.1 General Recommendations

The acoustics of worship spaces influence the way music is written and performed. Musical works are written for the reverberation characteristics of the places in which they are likely to be performed.

Organ music by Cesar Franck sounds significantly better in large cathedrals than some of the smaller scale organ music by Buxtehude and Bach, which, on the other hand, excels in small churches.

Sound systems for reverberation enhancements, particularly those outdoors, can be used to effectively simulate many types of spaces.

Some general recommendations for rooms intended for performances and the enjoyment of music in the worship space are:

- The background noise sound pressure levels must be low.
- The sound level must be optimized; it should neither be too loud nor too low. The sound level depends on the particular piece of music, the size of the ensemble, the way the music is performed, and the sound absorption that is due primarily to the worshippers.
- The spatial and temporal distribution of the early reflections must be good.
- The ratio between direct & early-reflected to reverberant sound must be appropriate to the desired clarity for the music that will be played.
- The reverberation time must be appropriate for the music that will be played.

The optimal reverberation time depends on the size of the room as well as on the type of performance that is expected to dominate the use of the room (see Figure 7.13 that shows some target values for various room volumes).

The worship space possibly can be designed to have variable acoustics by passive or by active means. Passive variation can be achieved using variable absorption. Sound systems can be used to adjust clarity and/or reverberation by active means.

A worship space that has a short reverberation time will be considered *dry and clinical* and acoustically uninteresting. Churches will typically require a ceiling height of approximately 10 to 15 m (30 to 45 ft) to obtain a sufficiently long reverberation time. One often strives for a frequency independent reverberation time within ± 5% in the octave bands from 250 Hz to 4 kHz.

Generally the sound absorption by common building constructions and surfaces is fairly low in the octave bands at 125 and below. Frequently, windows and poor floor constructions are the main sound absorbers at these low frequencies. It is common to allow an increase of reverberation time down to the 63 Hz octave band of approximately 50% over the value at 1 kHz. This practice results in good bass response and a warm-sounding reverberation.

The time gaps between the various components such as direct sound and major early reflections, are important and should be irregular and limited to a maximum of approximately 20 ms. The early reflections should be wideband, that is, not have a limited frequency range. One should avoid strong overhead reflections appearing in the same vertical plane as the direct sound. Because hearing cannot differentiate between signals having the same interaural delay time, such overhead reflections will cause comb filter effects unless masked by other reflections coming from the sides.

The spatial distribution of the early-reflected sound is important. Two effects are important, the *apparent source width* (ASW) and the *envelopment*, that is, the feeling of being immersed in reverberation arriving from all directions.

The apparent width of the sound source is determined by the way the early reflections reach the listener. The reflection angles should be close to the horizontal plane to maximize the advantages of our binaural hearing. One should also strive for many reflection incidence angles because this also contributes to the feeling of diffuseness in the sound field. It is advantageous if the directional distribution of the early reflections are such that when listening the sound field appears to be symmetrical. All of these requirements are easily fulfilled in conventional worship spaces such as in churches having cruciform or basilica floor plans.

To obtain good subjective diffusitivity, the reflections in and close to the median plane should be minimized relative to the lateral reflections. The side walls should be somewhat scattering, for example by using shelves, balconies, unevenly placed sound-absorbing patches, objects, etc., having sizes in the range of 0.5 to 0.05 m (1.5 ft to 2 in). Typically, windows, pillars, and other decorations such as large religious symbols (crucifixes, for example) will provide the diffusitivity.

One phenomenon of particular importance in long worship spaces is that of cancellation of direct sound on the main floor by the first reflection of sound off the audience in the frequency range between 125 Hz to 1 kHz—the seat dip effect discussed in conjunction with Figure 7.2. The cancellation will result in coloration of the sound, that is, the *direct* sound will sound weaker and thinner.

Because of the precedence effect, the direct sound determines our perception of the timbre of the sound source. Consequently, it is important to design the worship space so that coloration is minimized. This requires that speakers and performers are raised above the plane of the worshippers, that there are good sight lines for all worshippers, and that there are reflecting surfaces to fill in for some reduction in direct sound—if possible from the sides but if necessary from overhead.

It is also important that the acoustical conditions are such that they give musicians the possibility of hearing one another and give support for their own playing.

7.18.2 Organ Placement

Both the organ and the room in which it is placed are important for the quality of the organ sound. The size of the organ will depend on the desired organ loudness and that is determined by sound power, room volume, and room-sound absorption. The location will depend on visual and acoustical aspects and will involve balancing direct, early-reflected, and reverberant sound for desired clarity and reverberance (see Section 5.5). The Clarity metric C_{80} is frequently used to estimate the subjectively perceived clarity as described previously. Suitable clarity is critical for sound quality. Differences of a few dBs in C_{80} values may be easily heard. The desired clarity, and the associated values desired for C_{80} will differ depending on music type.

For the start transients of the pipe sounds to be heard, the organ speech needs to be clear. Even though the fundamental and the first overtones of flue pipes have long start transients (particularly at low frequencies), higher overtones and other pipe sounds (for example speaking noise), will have short transient times. Pipes such as reed pipes will also have short, initial transient times.

Reverberation is essential, however, for the fullness and mixing of orchestral sound as well as envelopment. Even though some mixing of the organ sound takes place in the organ case, a fairly long reverberation time (up to 3 to 4 s) is necessary for most religious musical works. The reverberation also contributes strongly to the celestial and numinous qualities of organ sound.

The organ serves to support both choir and congregational singing. For congregational singing, it is essential that the organ sound is similar for all congregants. This typically requires the organ to be placed well above the floor. If the organ is placed at floor level, the audience attenuation of sound will result in a large variation in the strength at different places (See Section 7.3). Additionally, placing the organ well above floor level reduces the risk of echo in churches that have substantial ceiling heights. It also preserves the clarity of organ sound because attenuation of direct and early-reflected sound due to the congregants is avoided.

Some possibilities for the placement of a pipe organ in a church that has a cruciform floor plan are shown in Figure 7.30. Similar possibilities also exist in churches having a basilica plan. In most churches, large or small, positions B and A1 in the figure will be preferred for sound quality and uniformity and will provide excellent organ sound for congregational singing, and will be suitable for the choir. In both cases, the organ needs to be well above floor level. Position B offers good reverberation with the desired celestial sound characteristic. For position B, the organ will be placed on a gallery in the back of the room (see Christ Church [Episcopal], Rochester, New York, Figure 7.31). Unless the room has extreme width or height, the side walls or ceiling will provide desired early-reflected sound. In Reform Synagogues, the organ typically will be placed at position B also.

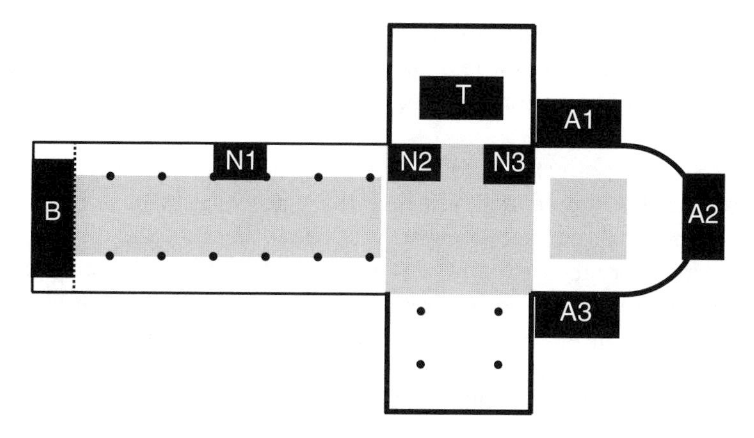

Figure 7.30 Some possible locations for the organ in cruciform Roman Catholic and Protestant churches.

Figure 7.31 The Craighead-Saunders organ on the purpose-built gallery in the Christ Church (Episcopal), Rochester, NY (organ maker and photo: Munetaka Yokota, GOart).

Position A1 has the advantage that it will give the best homogeneity of sound for the congregation, better than that of position B. The visual prominence may, however, be bothersome, and some prefer the organ not to be seen, but rather only heard as would be the case in position B.

Positions A2 and A3 are common in Episcopal and Anglican churches and are good for the choir and for reverberation. Sometimes, however, they result in insufficient clarity because of the design of the organ case and lack direct line-of-sight to the congregants.

Small churches are likely to have shorter reverberation times than large churches, but this should not influence the placement of the organ. In such churches, there may be excessive clarity of sound because the distances to the organ are small for all congregants and there is a lack of reverberant sound. But hiding the organ in a niche or transept (positions T, N2, and N3 in Figure 7.30) will greatly reduce its effectiveness in assisting the choir and the congregation. In small, nonreverberant churches, an electronic reverberation enhancement system may be used successfully (see discussion in Section 10.4.4).

Positions N2 and N3 are considered only for secondary organs, normally placed reasonably high for good sound distribution. They are useful positions when there is more than one organ or when the choir is in a nontypical position—seldom encountered in modern churches.

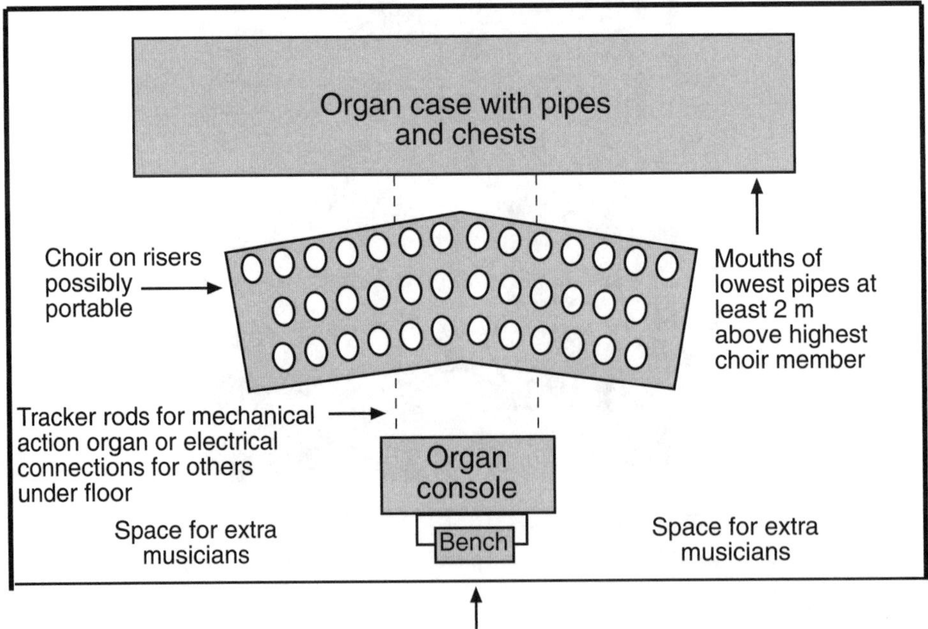

Figure 7.32 Suggested arrangement for organ and choir on top of gallery (adapted from Reference 7.8 and extended).

Figure 7.33 Organ on top of the rear gallery at North Shore Congregation Israel, Glencoe, IL (photo: David L. Klepper with Bolt Beranek and Newman).

Typically, balcony fronts will be reflective, but when the organ is placed on a gallery as in B, it is important that the gallery front is acoustically open (see Figure 7.33) so that the organ's sound is not hampered by a barrier.

The wall against which the organ is placed should be heavy and act as an efficient sound barrier in order to preserve the output power of the organ at low frequencies. Note also that it is important to take care in placing the blower for the organ so that its noise does not enter the worship space.

More information on organ placement can be found in Reference 7.10.

7.18.3 Organ and Choir Arrangements

In most churches music is frequently presented by the choir and organ together. This may be the choir singing to the congregation or the choir leading the congregation in singing. The most common arrangement is the choir at the front of the organ with the console located so that the organist can conduct the choir from the organ bench when a second musician is not present to conduct the choir. Such an arrangement is shown in Figure 7.32 (see Reference 7.8).

The mouths of the lowest pipes in the organ should be at least 2 m (7 ft) above the heads of the tallest choir members on the highest choir risers in order to prevent these choir members suffering hearing loss and being unable to coordinate with other choir members. At the same time, the organ sound should not be blocked from the choir and certainly not from the organist at the console. The rear gallery at the reform synagogue—North Shore Congregation Israel, Glencoe, Illinois shown in Figure 7.33 is a good example.

Here, the choir can be at the left or the right of the organ or one section on each side. They can hear the organ sufficiently for coordination, but not receive too high of organ sound pressure levels. The difficulty in that building is the location of the cantor, 30 meters away, at the front on the bimah, but a loudspeaker for coordination is provided in the cantor's lectern. Our understanding is that most cantors have preferred not to use it, but learn to anticipate sound from the choir and organ to keep synchronization for most of the congregation.

The stage of the Tanglewood Music Shed, Lenox, Massachusetts, presents still another possibility. Here, as shown in Figure 7.34, the organ is located above the reflective orchestral stage shell and can be heard in coordination with choir and the orchestra on the stage. The problem of coordination for stage musicians with the organ sound is solved by the openings in the stage ceiling that permit a moderate amount of organ sound energy to reach the stage musicians. The situation is similar at the Manhattan's Corpus Christi Roman Catholic Church, New York, New York, where the bottom surfaces of the organ's swell chamber and wind chests are among the actual overhead sound reflectors for the choir.

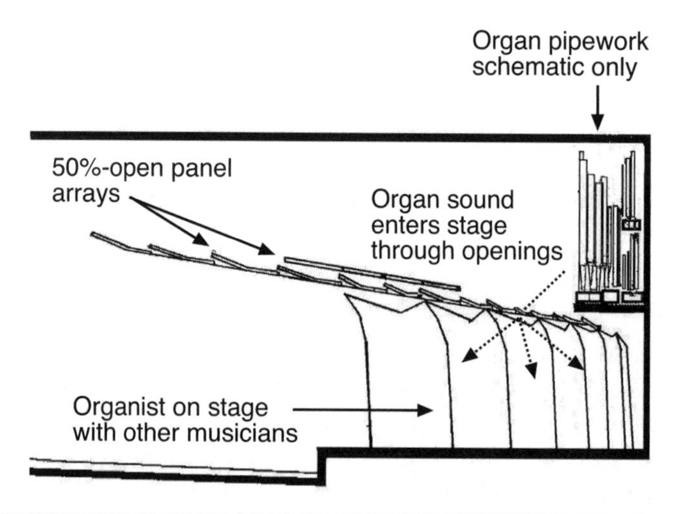

Figure 7.34 An example of how a choir and organ were coordinated at the stage of the Tanglewood Music Shed, Lenox, MA. It was the model for the solution at Corpus Christi Roman Catholic Church, New York, NY, near Columbia University, for the Holtkamp organ there. There the bottoms of the pipework wood air chests serve as the sound-reflecting panels with the open space between them allowing the right proportion of sound to reach the choir members' ears.

8

QUIET

8.1 NOISE, ANNOYANCE, AND SOUND QUALITY

A fundamental requirement of good room acoustics is the absence of not only such faults as echo, excessive, or absent reverberation, but also of noise and vibration.

The requirements for noise not to be damaging to hearing were discussed in Chapter 2. In designing the acoustics of rooms for worship, additional noise criteria must be met. The appropriate noise criterion will depend on the use of the room and on the requirement for visual and auditory impressions. We expect a concert hall to be quiet, but we will not be surprised to find an industrial workplace noisy. Sometimes noise is added intentionally as *acoustic perfume* such as Muzak-type background music, broadband noise having an unobtrusive spectral character, or even intentional heating, ventilation, and air conditioning (HVAC) noise.

The temporal and other characteristics of noise are also important as it relates to the annoyance level. One frequently uses various metrics of sound quality to further describe the properties of noise. The noise also may be described using other terminology, normally specific to a certain engineering field; for example, in describing the noise of air-handling equipment words such as hiss, rumble, and roar are used.

It is important to stress the need for low background sound pressure level (*SPL*). Most noise criteria have been set with persons in mind who have normal hearing. In societies with much sound-generating equipment (mechanical as well as electro-acoustical), people are subject to extra noise-induced hearing impairments. Persons who have hearing impairments or hearing loss generally have difficulty in understanding speech contaminated by noise and are often annoyed by noise even more than are persons who have normal hearing. Since most people develop age-related hearing loss (ARHL)—presbycusis—this is an important consideration.

The negative influence of HVAC noise on speech typically is due to masking. Frequency domain masking is primarily active upward in frequency. As indicated by Figure 5.3, it is reasonable to expect that the signal-to-noise ratio (assuming typical speech and HVAC noise spectra) must be at least 15 to 25 dB for speech intelligibility to be unaffected by noise in the speech-frequency range.

Some musical instruments such as the pipe organ may have support systems that generate noise. The blower needed for pipe organs frequently creates considerable noise that can be reduced by moving the blower out of the hall or church. However, the channels leading the air into the room may still radiate noise. This is more of a problem with older organs, particularly those from the 1890s to 1940s period. Modern organ builders do have techniques for *quiet winding* and, apparently, most organ builders in the Classic and Baroque periods developed techniques for quiet winding as well.

In large rooms for speech and music, particularly, the noise levels that can be tolerated are low. Whereas some noise may be tolerated during an actual service or performance because of the noise generated by the audience or worshippers themselves, there are always quiet moments in every worship service when obtrusive noise breaks the spell or spoils the mystery, and some music, including music for worship, actually demands fades to apparent, absolute silence.

Noise problems are generally a combination of characteristics of: (1) the sound source—mechanical equipment in most modern buildings; (2) the path—the kind of transmission problem discussed in Chapter 9 such as from a mechanical equipment room through a wall to a worship space, choir rehearsal room, or classroom, or perhaps through supply and return air ducts as discussed in this chapter; and (3) the receiver—the effect of the room that the listener is in and its own acoustical environment.

Specific problems can be solved by changes in any or all of these three elements, but the treatment of ductwork is usually the most practical. In addition, mechanical equipment causes vibration and this vibration can be transmitted through the building structure to critical spaces. Again, remedial measures may involve the source, the path, or the receiver, or any combination of the three.

Some environments are characterized by communication activities that are particularly sensitive to noise. This is particularly true for audio and video recording studios, radio and television stations, concert halls, operas, theaters, and similar venues. The background noise limits are usually set during consultation between the director, fundraiser, architect, and acoustician.

8.2 NOISE CRITERIA

Various space uses require different degrees of quiet. Significant experience and extensive subjective testing have provided a series of curves that show the allow-

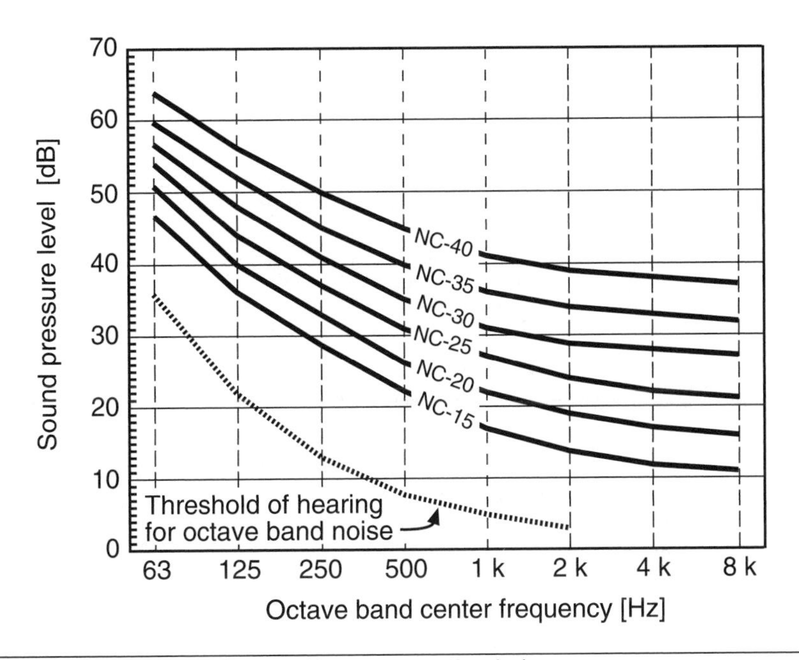

Figure 8.1 The *NC*-curves often used in room acoustics design.

able *SPL* in each octave band. The long-established *NC* curves (see Figure 8.1) are still recommended by the authors for most uses, whereas Beranek's room curves (see Figure 8.2) developed for concert hall and opera house applications and the new room criteria (*RC*) curves (see Figure 8.3) are recommended for worship spaces. Table 8.1 shows some typical *NC* requirements for various spaces.

The *NC* and *RC* curves are frequently used in determining whether a noise spectrum is acceptable or not, and also in specifying acceptable or maximum noise conditions. Such curves, specifying maximum octave band *SPLs* from 31 Hz to 16 kHz, are useful but do not take the temporal and cognitive character of noise into account.

In Europe, it is common to specify the maximum allowable noise level with dBA values. The U.S. way of specifying the maximum allowable noise as an *NC* or *RC* value and takes the spectral distribution of the noise into account is superior. The *NC* curves as well as the A-filter curve are based on auditory threshold data.

Noise shaped along the *NC* curves has a tendency to sound both *rumbly* and *hissy*. The curves were not shaped to have the best spectrum shape but rather to permit satisfactory speech communication without the noise being annoying.

The *RC* curves, on the other hand, have been shaped to be perceptually neutral; they are straight lines with a slope of −5 dB/octave. The *RC* method involves

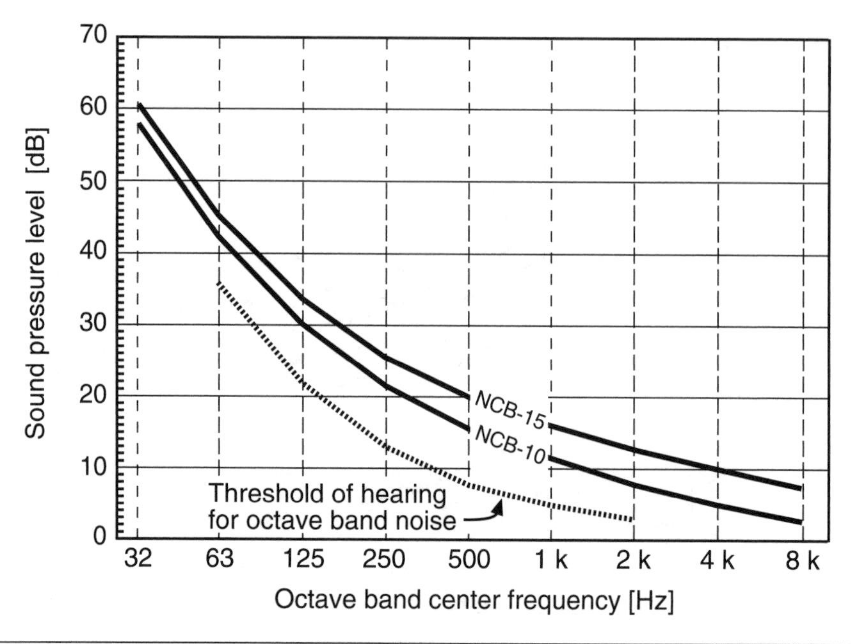

Figure 8.2 Beranek's balanced *NC*-curves.

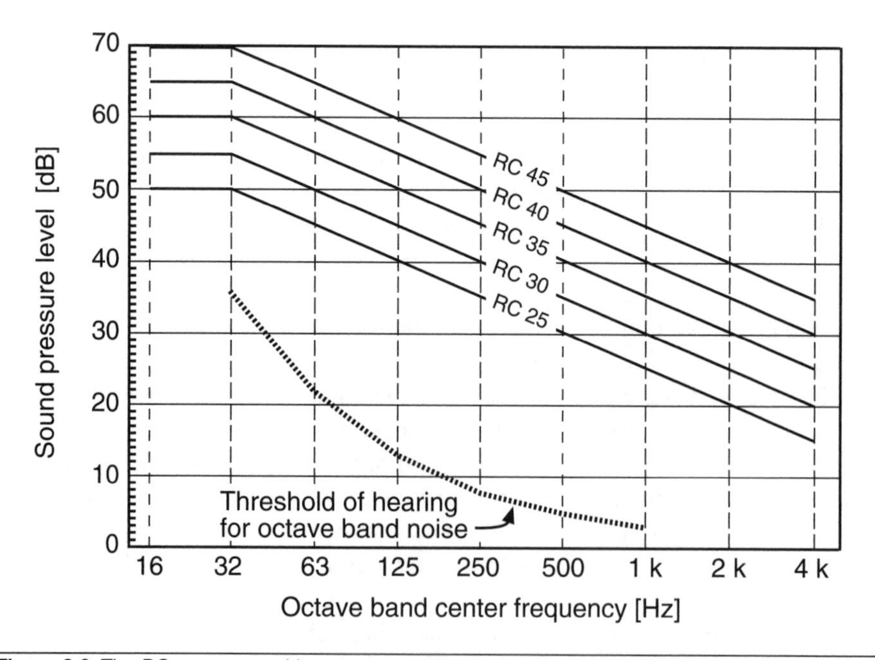

Figure 8.3 The *RC*-curves used in room acoustics design are better than the *NC*-curves for specifying the sound character and annoyance of room background noise. Note that there are several editions of these curves—these are the Mark II curves.

Table 8.1 Noise criteria for various spaces [1, 2, 3]. (The criteria use material from the references to this chapter but are somewhat modified by extensive experience of the authors with religious facilities.)

Worship spaces, theatres, opera houses, and concert or recital halls	NCB10–NCB15
Music and speech recording, radio, or TV studios	NCB10–NCB15
Music teaching classrooms; dependent on type of instruction	NCB15–NC20
Office and school auditoria; dependent on programmed uses	NCB10–NC25
House of worship: fellowship and social halls	NCB15–NC25
Board rooms, conference rooms, classrooms	NCB15–NC25
Motion picture theaters	NCB20–NC25
Libraries and study halls not used for worship	NC25–NC30
Private, enclosed offices	NC25–NC30
Open offices; "office landscapes"	NC30–NC40
Stores, coffeehouses, and restaurants	NC30–NC40

the determination of both an *RC* rating and a spectrum quality descriptor that determines if the spectrum is rumbly or hissy (see Reference 8.1).

In some cases one has to be content with a value for sound level in dBA. Concert halls typically require background noise sound levels well below 20 dBA.

The cognitive properties of noise are usually important in the environments under discussion here. It may be easier sometimes to tolerate a noise having a random character than a noise that is transient, that has a specific time pattern, or that carries information content. The spatial distribution of the noise is also important. A diffuse-sounding noise is usually less disturbing to some people than a noise that can be localized in space (although in the latter case one perhaps can more easily find and eliminate the noise source).

It is also important to avoid audible pure tones in the noise spectrum. For frequencies above 0.5 kHz, it is reasonable to require the tones to be inaudible which requires their *SPL* to be below the third octave-band level of the other noise in the band. In practice, this results in tones needing to be at least 5 dB below the level of the rest of the noise in an octave band.

8.3 MECHANICAL EQUIPMENT ROOM AND GENERAL ISOLATION

8.3.1 Basic Planning

The first rule of basic planning is to position mechanical equipment rooms as well as outdoor rooftop and ground level equipment as far as possible from sensitive spaces in which quiet is important. Buffer spaces—storage rooms, electrical switching rooms, possibly transformer rooms in which quiet transformers are installed, corridors, and lobbies—can be used to ones advantage. Generally, mechanical equipment rooms ideally should be installed in basements and a less desirable alternative would be on top floors or on roofs.

This thought process should be continued during all space planning. If a lavatory is located between a worship space and an exterior wall, all fixtures and plumbing should be attached to the exterior wall and not to the wall common with the worship space. Closets, store rooms, and lunch rooms can be located between worship spaces and mechanical equipment rooms.

8.3.2 Construction and Details

For penetrations of sound-isolating construction by pipes, conduit, and ducts, note the details of Figures 8.4a and 8.4b. In cases in which furring strips, glass fiber in the air space, resilient channels, and a second wythe of plaster and drywall or multiple drywall layers is applied to a basic masonry or concrete wall for double-wall sound isolation, the drywall or drywall and plaster layer should preferably be on the side away from the mechanical equipment room. The masonry will provide a better tie point for resilient support connections to pipes and conduit and certain mechanical equipment. (Chapter 9 discusses sound isolation and should be consulted for details regarding mechanical equipment room wall and ceiling construction.)

All duct elements should be designed to provide as smooth a flow as possible, avoiding abrupt changes in area or direction. Sound generated by air flow through 90° long-radius elbows, for instance, can be so low as to be negligible.

8.3.3 Doors

Anywhere in a building in which noise control is important, doors should be: as heavy as possible and gasketed around sides, top, and bottom; without grilles, undercut, or louvers; and be self-closing. When sensitive spaces must be located next to mechanical rooms, the mechanical room noise level must be studied on the basis of the known or predicted octave-band sound power level of all equipment. Additionally, a room-acoustics analysis (based on Chapter 2) and a complex sound-isolating construction plan (based on Chapter 9) needs to be completed. It is also crucial that two gasketed sound-control doors in a series, separated by 1 to 3 meters (3 to 10 ft), are installed for all passages between the mechanical room and the critical space.

8.3.4 Reverberant Sound Reduction

Generally, if all wall and ceiling surfaces in a mechanical equipment room are treated with 37 to 62 mm (1-1/2 to 2-1/2 in) thick glass fiber sound-absorbing

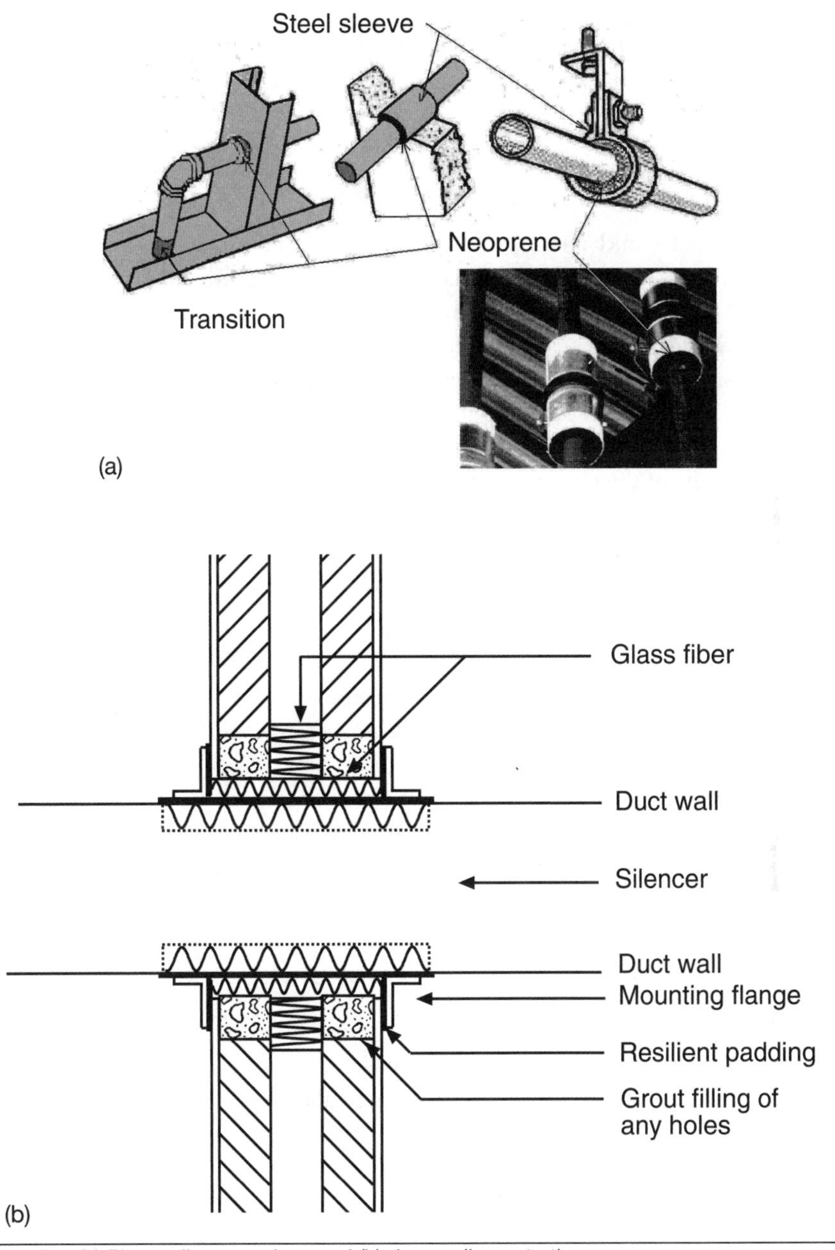

Figure 8.4 (a) Pipe wall penetrations and (b) duct wall penetration.

material, an approximately 6 to 7 dB reduction in mechanical equipment noise levels at 500 Hz and above can be obtained. The economics of this treatment compared with other construction noise-reduction measures should be analyzed for each case.

8.3.5 Mechanical Shafts and Chases

Mechanical shafts and chases should be treated the same way as mechanical equipment rooms, especially if they contain any noise-producing equipment even if noise-radiating ducts with thin walls are concentrated within them. Each shaft should be closed off at the mechanical equipment room and at each floor level, and wall constructions of the shaft walls themselves and the closures should be analyzed according to recommendations for sound isolation to adjacent spaces (see Chapter 9). Again, the principles presented in Figure 8.4 should be applied to the penetrations of the closure walls.

8.4 FAN SELECTION AND SOUND OUTPUT

8.4.1 Fan Types

The four basic types of fans/air-handlers are shown in Figure 8.5. There are trade-offs in efficiency, first cost, and noise level. Most quality manufacturers provide sound-output data in octave bands for near-capacity efficient operation, but sometimes data is lacking for operation at less than capacity, and this condition is usually noisier. Note that using a noisier but less expensive fan may result in more expensive noise control measures, making the overall installation more expensive. Long-term operating costs should, of course, always be considered.

8.4.2 Prediction of Fan Sound Power and Calculations for Room Noise Level

Manufacturers' data is preferred except when data is not available or when the design engineer has reason to question a particular manufacturer's testing proce-

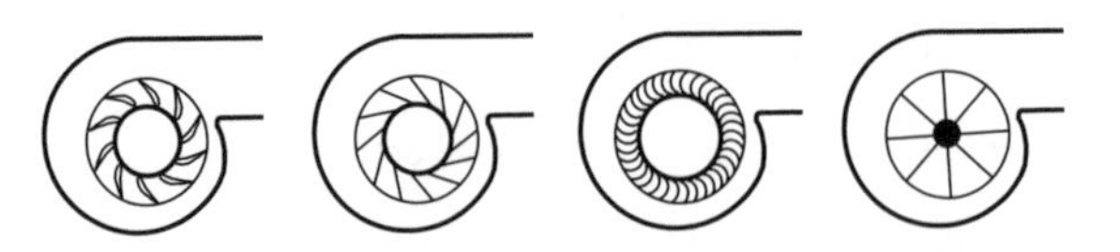

Figure 8.5 Fan blades: (from left to right) airfoil, backward-curved, forward-curved, and backward-inclined axial.

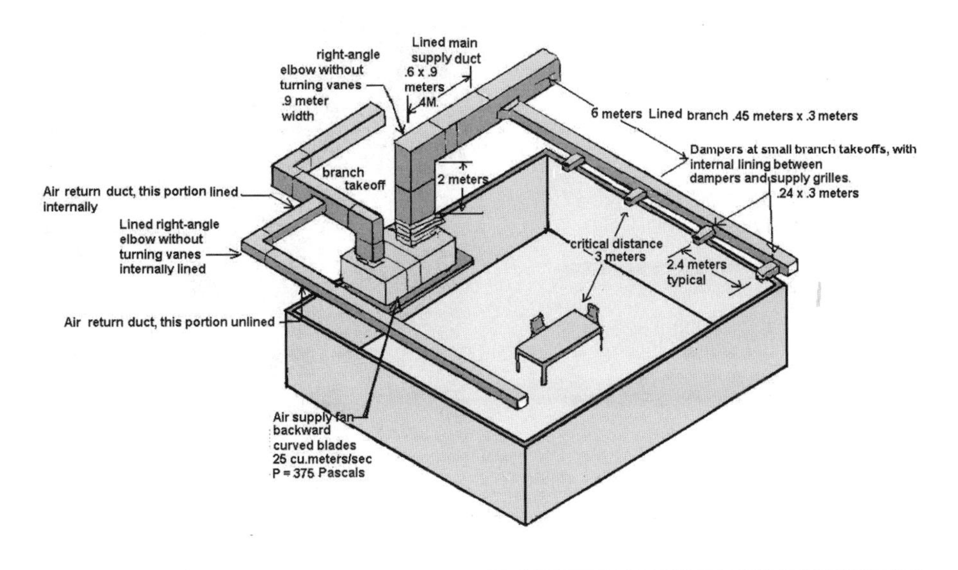

Figure 8.6 Elements of an office air conditioning system.

dure. Otherwise, the design engineer will need to estimate the sound power level in each octave band.

All sound power of the fan is not transmitted to the room because every element in the supply and return duct systems does absorb some sound energy. Figure 8.6 shows a typical system for a private office.

8.4.3 Silencers

If duct lining is insufficient to control duct-transmitted fan noise, or when ducts might otherwise negate sound isolation between rooms, silencers are employed. These are manufactured sound attenuators and a variety of designs are available (see Figures 8.7a and 8.7b). Custom designs to fit particular situations are also possible.

8.4.4 Diffusers, Grilles, and Dampers

The diffusers and grilles that provide termination for ductwork in a room provide *end-reflection* reduction of duct-born sound energy. The smaller the grille, the greater the end-reflection noise reduction (*NR*), and there is more reduction at low frequencies than at high frequencies. However, the grilles and diffusers are noisy elements in themselves because of the turbulence they introduce into the air stream. Grilles and diffusers can be selected to meet any design goal merely by keeping air velocity sufficiently low and avoiding any built-in dampers.

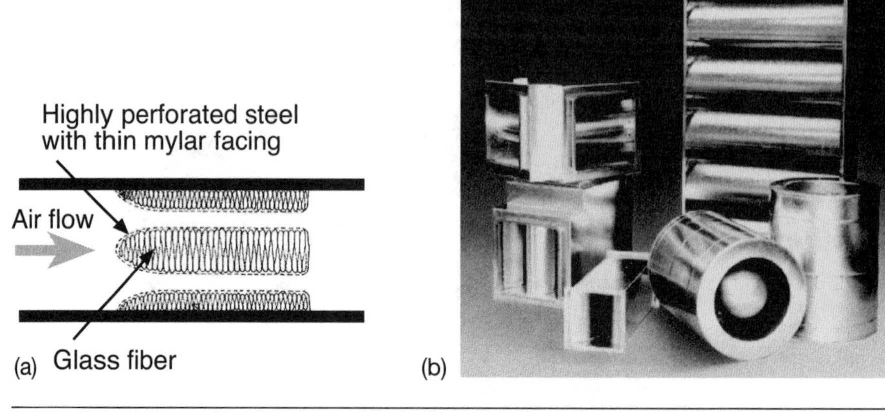

Figure 8.7 (a) Absorptive silencer and (b) commercial absorptive silencers (courtesy of Industrial Acoustics Co.).

For a given outlet style and air quantity, *SPLs* will increase approximately 2 dB with each 10% increase in maximum air flow velocity over the vanes and doubling the velocity will increase the sound level by approximately 16 dB.

Like fan manufacturers, grille manufacturers also provide sound ratings for their products. Selections can be based on manufactures' ratings but remember to account for the number of diffusers in a particular room; compare the dB addition chart in Figure 1.7 and the difference between room effects assumed by the manufacturer and those actually existing in a room. Proper installation and proper air flow conditions are also necessary. Control dampers that vary air supply also produce noise through turbulence and are better located upstream in supply ducts with lining between them and the grilles.

In worship spaces, concert halls, theaters, and recording studios, dampers should be separated from grilles and diffusers by at least 3 m (10 ft) of internally lined ductwork.

The kind of misalignment represented by the bad condition in Figure 8.8 can result in an unacceptable 10 dB increase in high-frequency *hiss* noise.

8.5 VIBRATION ISOLATION

8.5.1 Basic Planning

When minimum vibration isolation expense is required, large pieces of equipment should be floor-mounted on slabs on-grade. Where that is not possible, equip-

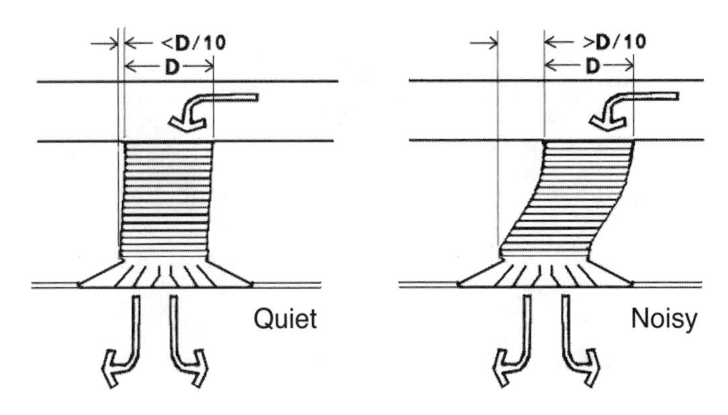

Figure 8.8 Good and bad alignment of grilles and diffusers.

ment should be located near bearing walls or columns, or over major beams. Hanging equipment from overhead major beams is acceptable, but lightweight floor slabs under critical spaces should be avoided in equipment installations.

8.5.2 Housekeeping Pads

Housekeeping pads are a 10 cm (4 in) thickening of the concrete floor under floor-mounted mechanical equipment. They are desirable because they add mass and stiffness where needed most and keep springs and other vibration-isolation devices off of the floor. Therefore, they add protection from rust and corrosion caused by draining equipment, minor leaks and spills, and regular floor washing. They may be poured as part of the floor slab or joined with steel rods and poured separately.

8.5.3 Floating Floors and Vibration-isolation Inertia Bases

Mechanical equipment may be placed on floating slabs unless it operates near the resonant frequency of the floating slab. Generally, this will be in the 7 to 15 Hz range and determined accurately by the springiness of the support material and the weight of the slab and the supported equipment, including water or other fluid contained in boilers and chillers. Figure 8.9 shows a typical mechanical equipment floated inertia base to give added weight and lower resonant frequency with much lower vibration amplitude.

Two common floated floor constructions are:

1. First the structural slab is inspected to insure it is airtight. Then, individual pads are spaced 0.3 to 0.6 m (1 to 2 ft) on centers each way and covered with plywood or sheet metal. They are then waterproofed,

(a)

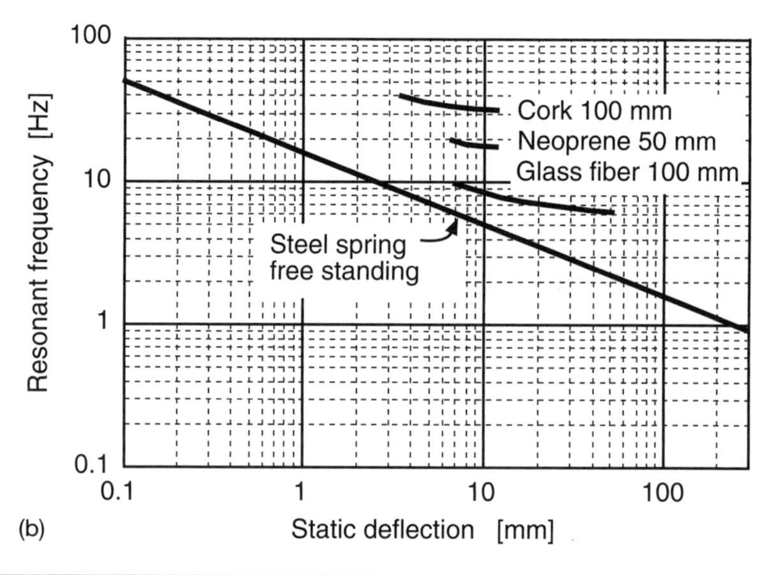

Figure 8.9 (a) Use of a Kinetics concrete inertia base T-shape form with 1 inch deflection seismically restrained isolators. The inertia base uses a T-shape design to support the water weight in the inlet and discharge elbows. The isolators are mounted to oversized base plates to help distribute the seismic load from the isolators to the structure (photo: Kinetics Noise Control). (b) Range of application for different types and the use of a concrete inertia base to reduce vibration amplitude.

often with plastic coating. In some cases additional isolating is added by placing light-weight waterproofed glass fiber blankets in the spaces between the pads to reduce reverberant sound buildup in this space. Approximately 5 or 6 dB of additional sound attenuation is all that can be expected from the addition of the glass fiber. Reinforced concrete is then poured directly on the waterproofed plywood or sheet metal, or precast concrete slabs are used and the joints between them also sealed airtight.

2. The second system differs from the first in that cast-in place canisters are installed on centers each way, 0.6 to 1.2 m (2 to 4 ft) with the second reinforced concrete layer poured over the reinforcing rods and widened bases of the canisters up to the top of the canisters or over them with holes left to reach the canisters. The canisters consist of springs prepossessed by jacking screws. After the second layer of concrete has cured, these screws are removed or loosened with the entire second layer raised to float, any holes grouted, and the equipment installed.

8.5.4 Pipe and Duct Connections

Connections should be resilient so that pipes and ducts do not short circuit the isolation of the vibrating equipment to the building structure. Materials include

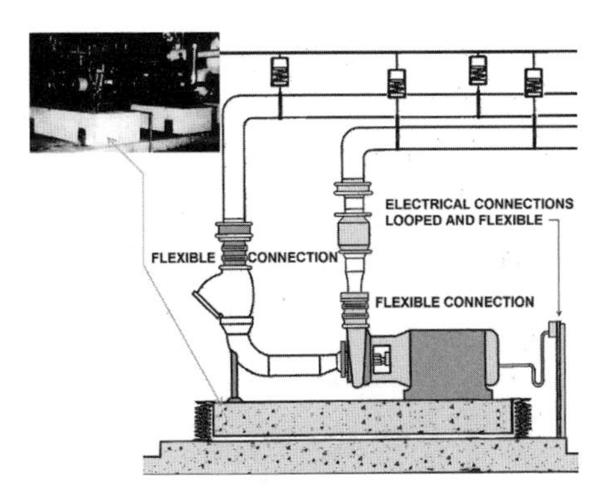

Figure 8.10 Vibration isolation of a pump, including pipes and conduit, similar to that shown in Figure 8.9a.

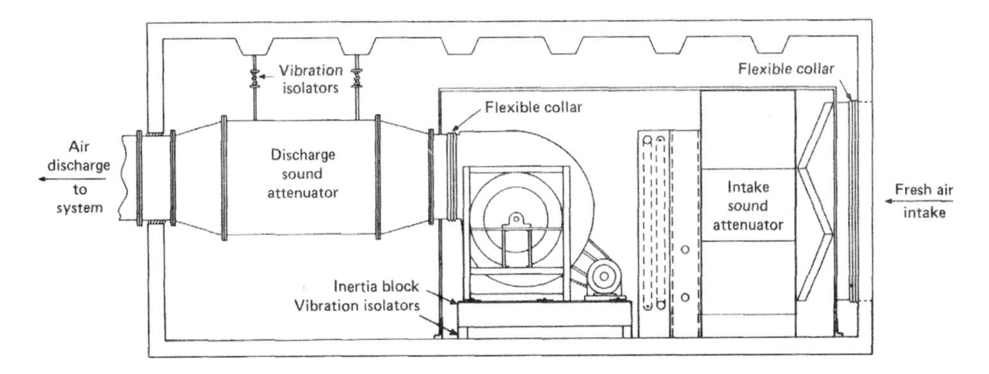

Figure 8.11 Isolation of a fan. Electrical connections handled the same as that in Figure 8.10 (see Reference 8.4).

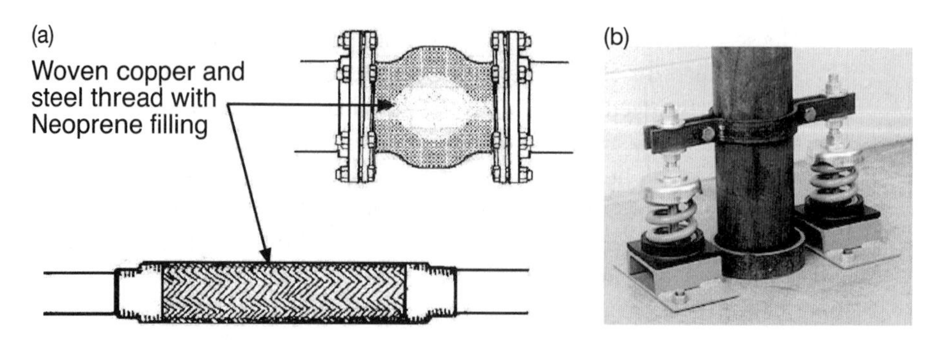

Figure 8.12 (a) Examples of flexible pipe joints. (b) Steel springs used for vibration isolation of pipe riser (photo: Kinetics Noise Control).

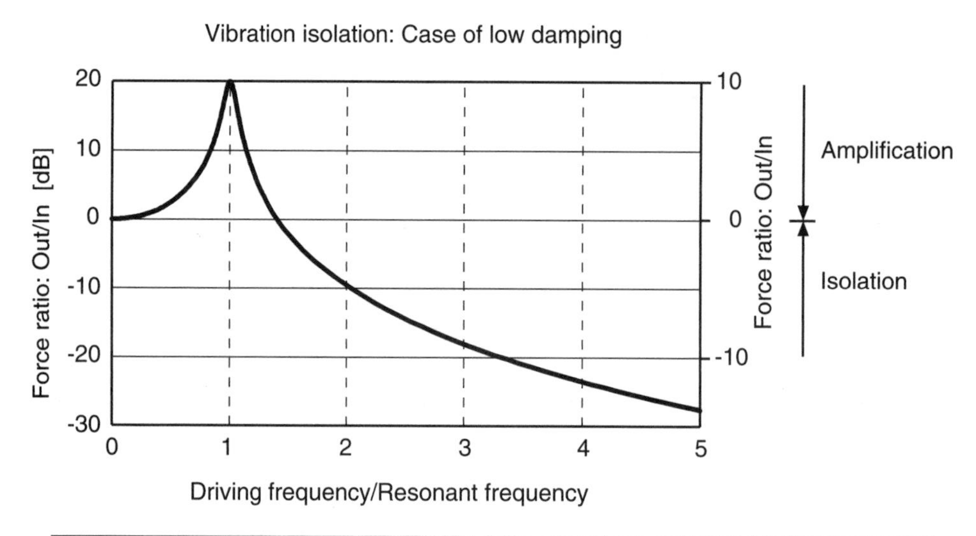

Figure 8.13 Design graph for isolator requirements is shown for a case of low damping such as when using steel springs, assuming a stable structure under the spring. Modifications and additional care is required if the structure is weak. Note that improper selection of vibration isolation mounts can not only result in less-than-adequate isolation, but can actually result in amplification of vibration by resonance. The resonant frequency is determined by the isolating spring and the mass it supports.

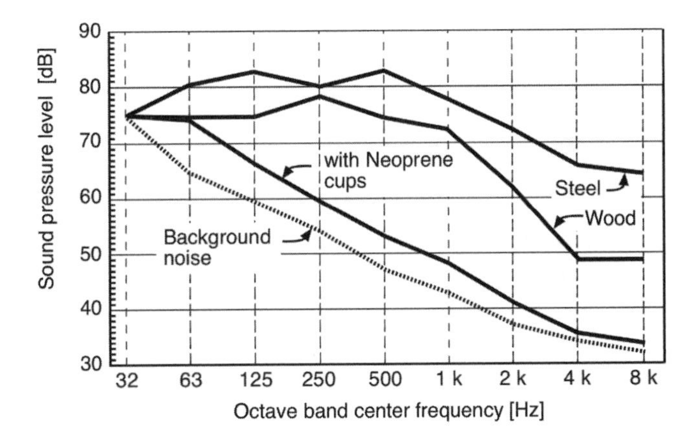

Figure 8.14 Reduction of chair-scrape noise by the addition of neoprene rubber cups. Chairs with steel and wood legs respectively.

heavy canvas, neoprene or other synthetic rubber, and woven metal combined with these materials. See Figures 8.10 through 8.13.

8.5.5 Chair-scrape Noise

The problem of metal-legged chairs moved on slate or ceramic floors is discussed with reference to synagogues, but it can be a problem in many worship and study buildings. Synthetic rubber cups applied to the bottoms can solve the problem. Figure 8.14 presents some data obtained at Yeshivat Beit Orot, Jerusalem. Hardware stores can provide several sizes to accommodate most cases.

9

SOUND ISOLATION AND OTHER NOISE ISSUES

9.1 SOUND TRANSMISSION

Sound may be transmitted from room to room through an intervening wall; *direct transmission* along an interconnecting structure as vibration, *flanking transmission*, or through open-air paths as shown in Figure 9.1. Sound will be attenuated while travelling along any path, but even small holes can make a costly sound-isolating wall ineffective. Sound can travel great distances as vibration with little attenuation along light, continuous curtain walls, and lightweight suspended ceilings, making flanking transmission a serious problem.

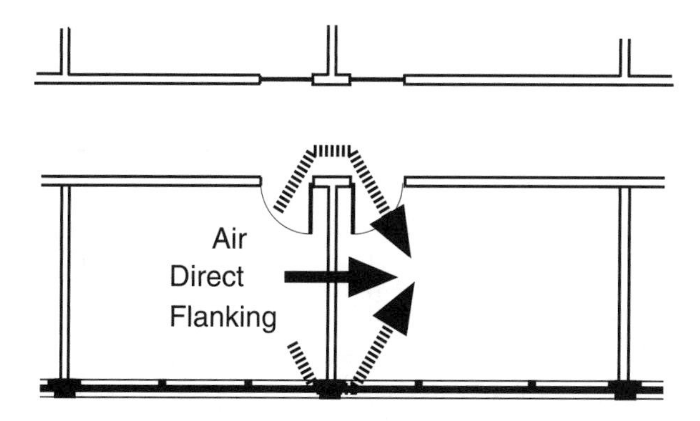

Figure 9.1 Three types of room-room transmission paths: direct, air, and flanking.

9.2 NOISE REDUCTION AND SOUND TRANSMISSION LOSS

We discuss two basic acoustical quantities or concepts regarding sound isolation problems: (1) noise reduction (*NR*) and (2) sound transmission loss (*TL*). Values for both are expressed in decibels.

Let us assume that we have two adjacent rooms without interconnecting air paths and that the flanking structure-born sound transmission paths are of minimal importance as illustrated in Figure 9.2. In this example, sound can be significantly transmitted, only through the intervening wall. Further, let us assume we generate a steady sound in the first room (source room) at a sound pressure level of SPL_1. In the second room (receiving room), we find that the sound is reduced to level SPL_2. The *NR* between the rooms is the arithmetic difference *SPL*.

$$NR = SPL_1 - SPL_2$$

The *NR* is independent of source room level but is dependent on three variables:

1. The area of the intervening wall in m² (S)
2. The absorption area of the receiving room in metric Sabins (A_2)
3. The *TL* of the intervening wall

The relationship of these quantities is illustrated in the following equation:

$$NR = TL - 10\log(S/A_2) = TL - RC$$

Normally, the *TL* will be much greater than the room correction (*RC*). In practical situations, it may range from 10 to 90 dB, but *RC* seldom varies over a range greater than −6 to +6 dB. A level difference of 6 dB is significant. Note that if $S = A_2$, *RC* will be 0 dB and, thus $NR = TL$. There are many situations in which this is the case.

The *TL* for any type of wall (for example, 8 in or 200 mm brick) is theoretically the same regardless of the location on the wall area or the dimensions of the

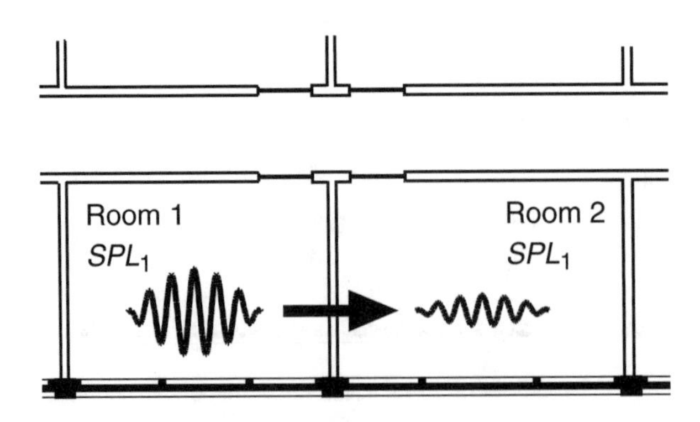

Figure 9.2 The wall is difficult for the sound wave to move, hence there is a room-to-room blockage of sound transmission.

wall and its resulting area. The *TL* depends on the wall cross section. We can select walls or design for sound isolation using established *TL* values for various walls. However, we discover what the total *NR* will be, including the effect of the wall area and the absorption in the receiving room. *Our ears will ultimately sense the NR but not the TL.* We might select a wall with a *TL* of 40 dB. Between two reverberant rooms with a large intervening wall area, the resulting *NR* might be 35 dB. Between two *dead* or sound-absorbing rooms with a small wall between them, the *NR* might be 45 dB. This 10 dB difference in performance would be important.

The sound absorption in the receiving room is important to the resulting *NR* as illustrated in Figure 9.3. The *RC* will be different in these two cases, leading to

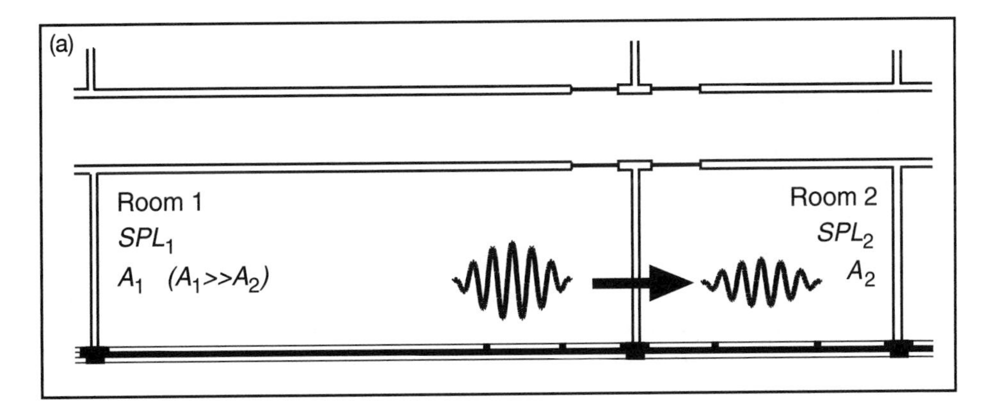

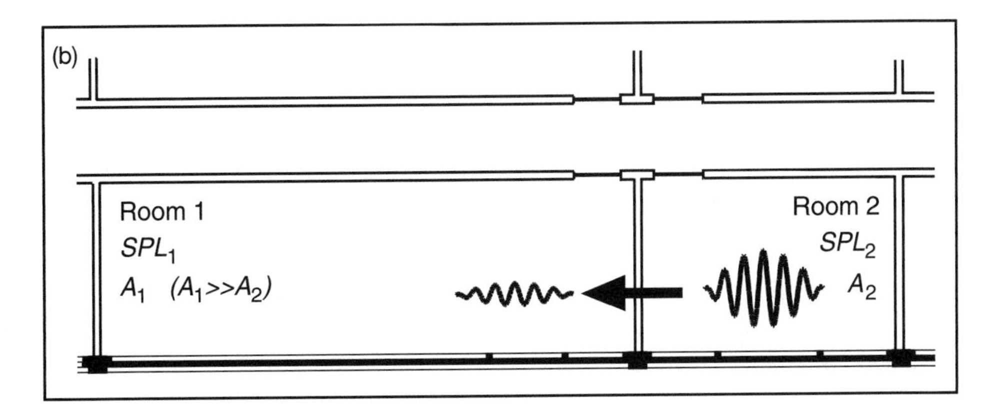

Figure 9.3 Two cases of sound isolation. (a) The receiving room has a small sound-absorption area A_2. (b) The receiving room has a large sound-absorption area A_1. The wall size, *S*, is the same in both cases.

higher *NR* in the case on the bottom than in the case on the top $A_1 > A_2$. The *RC* will be:

$$10 \log(S/A_2) < 10 \log(S/A_1)$$

It is important to remember that in theory the *TL* for any given wall is always the same, but in fact it will vary with *edge conditions*, the nature of the joint between the wall and adjacent structures, as well as a slight variance dependent on area. The *TL* will definitely vary with workmanship. The better the workmanship, the better the *TL* and its closeness to the predicted values in theory.

9.3 EFFICIENT SOUND ISOLATION

Good sound isolation (or *NR*) or high sound *TL* is provided by massive, limp, and impervious material. The heavier the material the better. Concrete that is 150 mm (6 in) is better than a 1 mm steel sheet. A steel sheet, however, is superior to a plywood panel of equal mass because the steel is less rigid.

Porous sound-absorbing materials are extremely poor isolators because of the open pores and the low mass of the porous sound-absorbing sheet. Sound-isolating materials must not transmit air (it must be impervious) and must either be rigid or have high mass per unit area. It is important to note that heat-insulating materials are generally poor sound isolators. One should not confuse the NR provided by acoustic tile within a room with *NR* between rooms.

9.4 EFFECT OF BACKGROUND OR MASKING NOISE

In solving a sound isolation problem, the most important elements are:

1. The source room (Room 1) sound pressure level (SPL_1)
2. The *NR* due to the wall and the sound-absorption area in Room 2 that contribute to the resulting receiving room sound pressure level (SPL_2)
3. The steady background noise in the receiving room (SPL_B).

If a *NR* is designed to reduce source room level SPL_1 to receiving room level SPL_2 equal or below the receiving room background level SPL_B, the goal has been obtained (see Figure 9.4). If the sound level SPL_B is 10 dB above SPL_2, the sound from the source in Room 1 will usually not be audible in Room 2. (Assuming that the noise characteristics are not too dissimilar.)

A quiet building in the country will require better sound-isolating construction than a moderately noisy, ventilated building in the city.

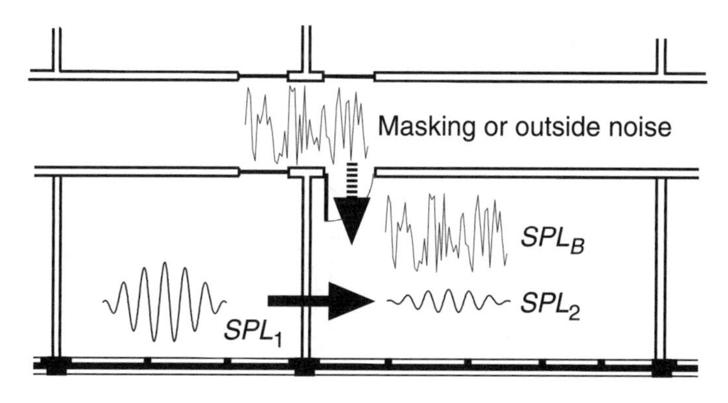

Figure 9.4 Sound isolation with masking noise.

9.5 TRANSMISSION LOSS CURVES OF TYPICAL WALLS

Like all other acoustical problems, sound isolation must be considered with respect to frequency. The source and receiving room sound pressure levels, SPL_1 and SPL_2, will vary with frequency and so will the *TL* of a wall and the absorption area A_2 of the receiving room. Thus, designs for sound isolation must be *spectrum* designs, considering each octave band in the frequency range of interest, not average number designs. In practice, we must consider octave-band *NR* results from 125 Hz to 4000 Hz for speech sound isolation, but additional bass octaves must be considered for music.

A perfectly limp wall will have a *TL* curve that rises at a rate of 6 dB per octave (i.e., 6 dB per doubling of frequency) as shown in Figure 9.5. Also, a perfectly limp wall that is double the weight of another perfectly limp wall will have a *TL* of 6 dB greater than the lighter wall at any frequency. Actually, a perfectly limp wall is rare; nearly all walls have some stiffness.

Consider the situation shown in Figure 9.4. If the sound transmitted through the wall is a speech signal, having a sound pressure level SPL_2, that is, SPL_B in the Equation 9.3—the *SPL* due to the masking noise—normally, the transmitted speech will be slightly audible but not intelligible. With this design goal, we have the design criterion:

$$NR = SPL_1 - SPL_B$$

One can vary the *NR* by varying the *TL* of the wall construction in respect to its area, taking care with edge effects as well as by increasing the receiving room absorption, A_2. From the viewpoint of privacy, the influence of a steady background noise is always important; the *SPL* due to the masking noise is just as important as the *NR*.

In many rooms, other than worship, concert, lecture, and theatre spaces, 10 dB more background noise can be achieved more economically than 10 dB more *NR*, and yet the background noise will still be reasonably unobtrusive for the use

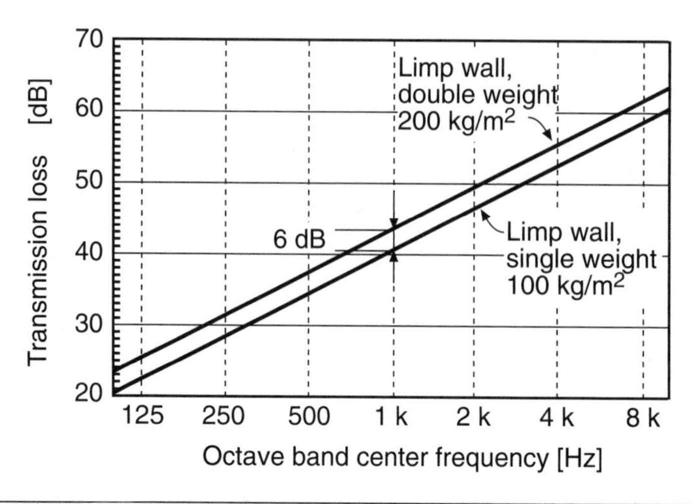

Figure 9.5 *TL* curves for two limp walls that have different mass per unit area.

of the space. For example, one might generate moderate air flow noise with a diffuser in the air outlet in the receiving room. Libraries and offices are particularly good spaces for this technique.

9.6 A BALANCED SPECTRUM DESIGN

The diagram in Figure 9.6 is an example of a balanced spectrum design approach to sound isolation. Note that the *NR* values typically increase with increasing

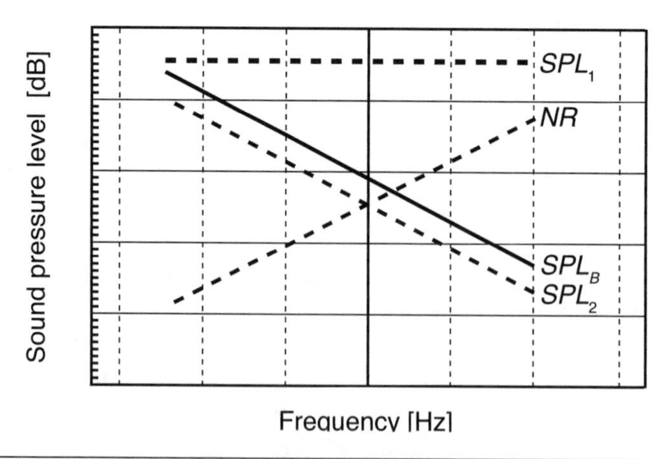

Figure 9.6 Spectrally optimized design approach to privacy. The sound level of the masking noise should be at least that of the transmitted noise at all frequencies.

frequency. The sound level of the masking noise SPL_B typically decreases with increasing frequency. However, many noise sources such as radios, televisions, and musical instruments, for example, have spectra that are not flat. They also have recognizable sound features, making them irritating to listen to even if they barely *break through*. For such noise sources, it is advantageous that the transmitted sound level is at least 10 dB below the masking noise. The spectral shape of the masking noise will depend on its source; traffic noise (transmitted through windows) and ventilation noise (for example, air conditioning) have different spectral shapes.

9.7 THE COINCIDENCE EFFECT

Any real wall will have stiffness along with mass. Because of the stiffness, there will be resonances at which the wall will prefer to vibrate. The *TL* at these resonance frequencies will be much lower than that predicted by the limp-wall mass straight ascending line shown in Figure 9.5. One of these effects is called the coincidence effect. At and around the coincidence frequency, the wall will be nearly transparent to sound at certain angles. The incoming sound will drive the wall, creating a flexural wave in the wall as shown in Figure 9.7. The flexural wave will radiate sound on the other side of the wall.

Figure 9.8 illustrates the characteristic behavior of the *TL* of a wall that has stiffness. Well below the critical frequency at which the coincidence resonance starts for an infinite wall, the *TL* increases by +6 dB per octave. In the coincidence region, it is inversely controlled by stiffness and the internal damping of the material controls the plateau frequency spread. A plywood panel contains considerable internal damping. Glass and steel have relatively less internal damping. A material that has little damping gives off a sound such as steel rings do when tapped. A material that has high damping will give off a thud.

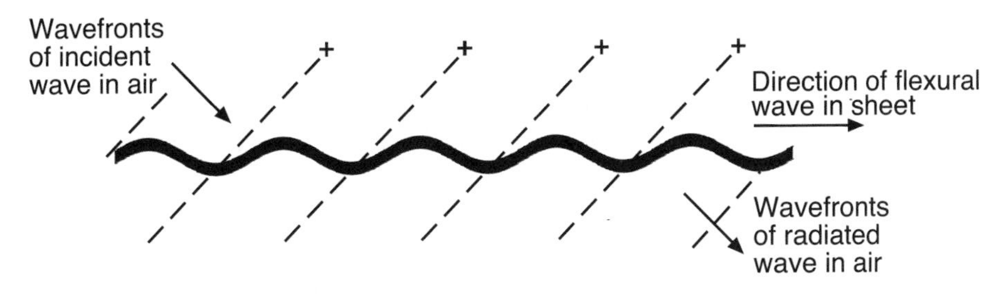

Figure 9.7 The incident sound wave forces its vibration pattern on the wall, creating a flexural wave that radiates sound on the wall's secondary side.

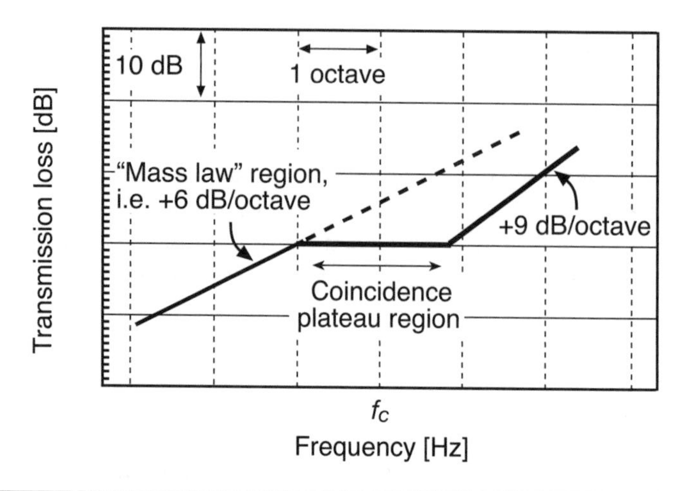

Figure 9.8 Outline of the characteristic behavior of the *TL* of a wall that has stiffness compared to a limp wall. The height of the plateau is inversely controlled by stiffness. Damping controls the plateau frequency spread.

The frequency at which the critical frequency occurs is decided by the mass per unit area of the wall and of its stiffness. With the same materials and varying surface weight, the *TL* curve slides one octave up for halving of surface weight and one octave down for doubling surface weight. The plateau height and width stay the same and the entire *TL* curve moves horizontally along the frequency scale.

The net effect of doubling surface weight is a 6 dB increase in *TL* except around the plateau where it remains the same. There are more economical ways of increasing the *TL* by 6 dB than by doubling the wall thickness as will be shown later. (Doubling the thickness of a 37 mm drywall wall is economical; doubling the thickness of a 200 mm block wall is not.)

9.8 AVERAGE TRANSMISSION LOSS AND SOUND TRANSMISSION CLASS

Single number ratings for sound isolation are published for various wall, floor, and ceiling constructions and, if single-number analysis must be used, such values are more meaningful than averages because it derates a construction for serious dips and plateaus in the *TL* curve. This may have minimal effect on the average but a significant effect on the degree of sound isolation that is ultimately the design goal. Two walls that have the same average *TL* but in which one curve exhibits a serious *TL* dip will be rated differently; the one with the dip will carry a lower rating. Music isolation demands complete octave-band analysis.

Two types of single-value ratings of sound *TL* are the sound transmission class (*STC*) rating used in the United States and the sound reduction index (R_w) used in Europe. These are defined by national and international standards. The definition is usually modified nationally so one should check the national standards that are applicable.

9.9 LABORATORY VS. FIELD MEASUREMENTS

Many publications report *TL* values. One should look at all such data skeptically. Three types of measurements are usually reported:

1. Field measurements from actual partitions installed in buildings—field sound transmission class (*FSTC*)
2. Laboratory measurements on large construction samples built in a field-like manner (*STC*)
3. Laboratory measurements on carefully constructed small samples.

The range of data variation between such measurements for the same partition construction can be as much as 10 dB from the field to the small laboratory sample. Edge conditions have an influence as do field inaccuracies. One should always seek data based on measurements of large samples—at least 2.5 by 3 m (8 by 10 ft)—built as one would expect in the field.

Light metal studs, particularly trussed wire studs, are preferable to wood stud walls. Heavy, sand aggregate plaster is preferable to lightweight vermiculite and perlite aggregate plaster. Plaster of gypsum lath is stiffer than plaster on metal lath and is thus a better sound isolator. Stud walls using laminated gypsum board facings (drywall) compare favorably with wet plaster, but one must be careful of leaky joints where gypsum board abuts concrete floors, ceilings, and columns. Staggered stud construction (separate studs for the support of each side) and the use of resilient channels or clips between studs and facings can add as much as 10 dB to a wall's isolation, particularly if combined with some inexpensive glass fiber blanket in the air space. Glass fiber is not particularly effective if the two sides are tied together rigidly.

9.10 EFFECTS OF LEAKS

Leaks in partitions, for example between a gypsum wall and a concrete floor, are frequently covered by wooden or plastic trim in the field so that it is not visible. Figure 9.9 shows a serious leak behind wooden trim. Trim has virtually no influence on the acoustical behavior of the leak unless caulking material is used. Movable partitions and doors are particularly prone to leakage.

Figure 9.9 A major leak between a lightweight concrete wall and the floor that was hidden by wood trim.

Because leaks have virtually no *TL* at any frequency, even a small leak can seriously degrade an otherwise good wall. Building an expensive wall and tolerating leaks created during construction is a waste of money. A leak having an area of 1 cm^2 in a 10 m^2 wall will reduce the obtainable *TL* to a maximum of approximately 50 dB irrespective of the *TL* of the wall itself.

9.11 COMPLEX OR DOUBLE PARTITIONS

Complex or double walls are efficient ways of achieving lightweight walls that have high *TL*. Such partitions have two or more sheets of material such as plasterboard or plywood separated by an air space. The air space acts as a spring that mechanically decouples the two sheets. The spring and the mass of the sheets cause a low-frequency resonance that leads to low *TL* at the resonance frequency as presented in Figure 9.10.

The increase of 12 dB per octave in *TL* is a theoretical value that cannot be reached in practice because of mechanical coupling between the sheets and resonances in the air space between the sheets. This can be eliminated by filling the air space with a porous absorber and by using separate studs for the two sheets. Because the sheets typically have a critical frequency of approximately 2.5 kHz, the *TL* will ultimately be limited by the coincidence effect and the associated coincidence plateau. This effect can be offset slightly by using sheets that have different thicknesses. Figure 9.11 shows the *TL* curves for some double walls compared to that of a 160 mm concrete wall. One notes that at low frequencies the concrete wall is much superior.

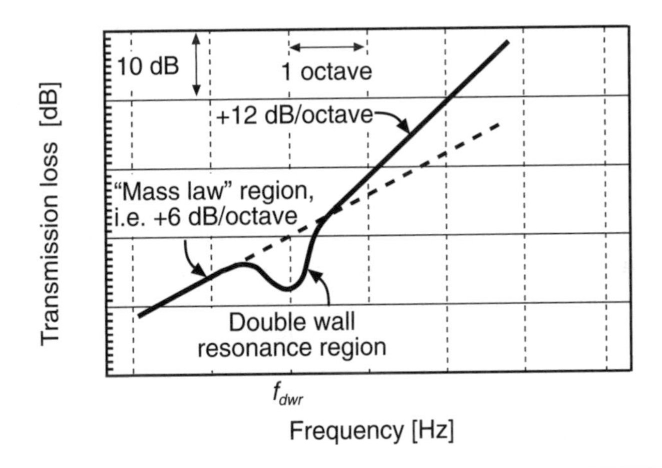

Figure 9.10 The *TL* of a double-panel partition (solid line) as a function of frequency. The *TL* of a single-panel partition (dashed line) having the same total mass as the double wall shown for comparison. Both curves shown for $f_{dwr} << f_c$.

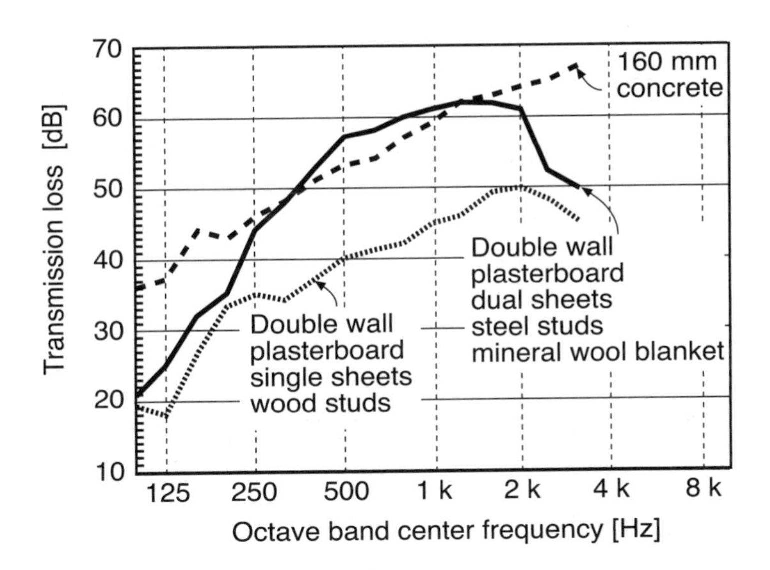

Figure 9.11 Some *TL* curves for two double walls compared to that of a 160 mm concrete wall. Note the influence of the double-wall resonance and the coincidence on the *TL* of the double walls.

A special type of double wall is the *resilient skin* used to enhance the sound isolation of heavy walls even more. A resilient skin is typically a sheet of plasterboard attached to the wall by 50 mm studs or resilient clips and glass wool filling the air space. Figure 9.12 illustrates the 10 dB effect of such a resilient skin on the *TL* of a brick wall 200 mm thick.

Multilayer partitions characterized by high *TL* are a floating floor combined with a suspended ceiling as given in Figure 9.13. Figure 9.14 shows a heavy wall that has been further improved by the addition of an added resilient skin. The constructions shown in these two figures can serve as a model for cases in which worship spaces are located directly above classrooms or a social hall.

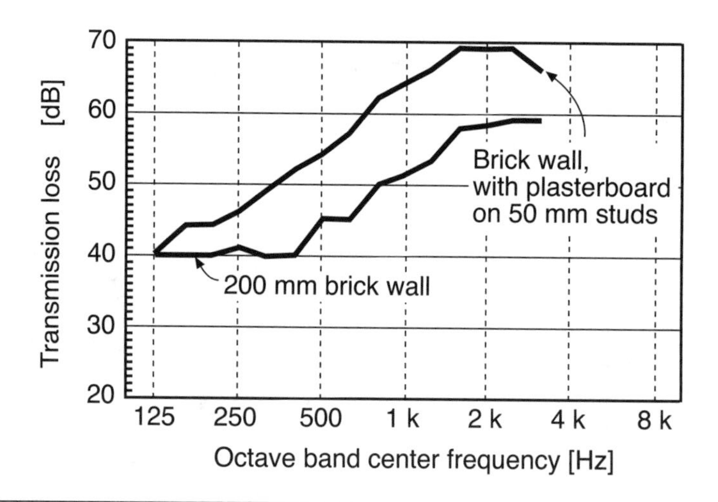

Figure 9.12 Example of the *TL* of a brick wall 20 cm thick, with and without a resilient skin. In this case, the resilient skin consisted of a plasterboard sheet 13 mm thick on 50 mm studs. The air space was filled with a 25 mm mineral wool blanket (see Figure 9.14).

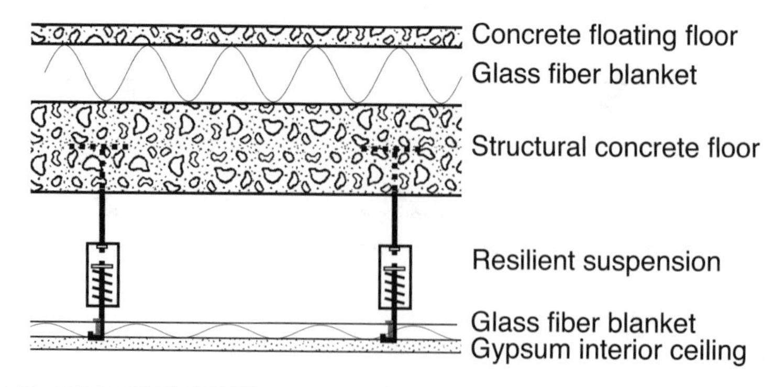

Figure 9.13 Example of a multilayer floor with high *TL*.

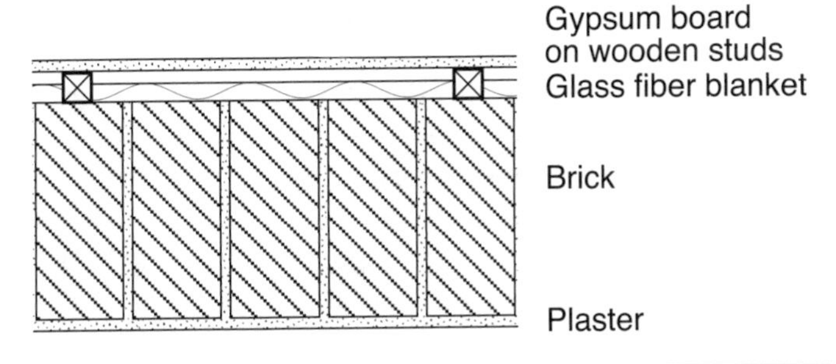

Figure 9.14 Example of a brick wall with a resilient skin characterized by high *TL*.

9.12 CONTROL OF FLANKING SOUND TRANSMISSION

Flanking sound transmission is transmission around a barrier via any number of paths. Less sound energy actually might be transmitted through a wall than through flanking transmission via a curtain wall, continuous lightweight ceiling, interconnecting ductwork, or doors and a corridor. For normal, good-quality construction, one cannot expect more than 50 to 55 dB of overall *NR* between adjacent rooms unless interconnecting structure transmission is eliminated by resilient airtight joints or box-within-a-box construction.

The inner box must have full resilient separation from the outer box without any rigid ties. To exceed 55 dB *NR*, special corridor doors or double doors will be required and special care taken with interior treatment or silencers in interconnecting ducts.

Some buildings lack forced ventilation and, therefore, have minimal background noise, however, construction required for little background noise may be too expensive. Open windows, louvered and undercut doors, and hard floor finishes with minimal impact noise isolation are other typical problems in such buildings.

Airtight isolation in movable partitions in today's flexible buildings requires extreme care. Care with gaskets as well as compensation for unexpected and expected construction tolerances and structural deflections are all required. One should select gaskets that are of an airtight, durable material and that are soft. Care with gaskets is particularly important with partial height partitions that only reach a suspended ceiling instead of reaching the construction above. Sound can be transmitted through the ceiling of one room, through the open plenum, and down through the ceiling of the next room. This path should be checked to insure higher overall *TL* than the wall. Large return openings into the plenum space above a ceiling can be a problem. Masking noise, either from the ventilating system or electronically generated, is mandatory in offices with movable partitions. This also will be necessary when movable doors are used to provide flexible space.

9.13 MUSIC PRACTICE AND TEACHING FACILITIES

In planning the acoustical environment of music teaching facilities, two overriding concerns (apart from obvious room acoustics considerations) are the sound isolation of critical spaces and mechanical equipment noise and vibration control, including control of crosstalk, which is the flanking transmission between adjacent spaces referred to earlier. Other potential flanking paths in a typical music building include doors, pipe penetrations, electrical outlets, and windows. The cumulative effect of all these potential sound paths must be anticipated and controlled.

9.13.1 Basic Planning

Critical music teaching and rehearsal spaces requiring high levels of sound isolation should be arranged to take advantage of corridors, storerooms, and similar noncritical spaces as buffers or sound locks between such spaces. Where this is not possible, one should group music spaces with potentially high live instrumental sound levels together and not intersperse them with less active classrooms and lecture rooms used primarily for speaking. One can also save on costs by locating spaces with high sound-isolation requirements on-grade, where heavy masonry cores of partitions and where special floated floors are not necessary. Mechanical equipment rooms should be as far away from the most sensitive spaces as possible. In an urban environment, one should locate the mechanical equipment rooms on the side of the building with the greatest exposure to exterior noise such as heavy street traffic. Pumps, compressors, and chillers should be located on-grade. Single-story construction for equivalent space is always less expensive than multistory and the fewer the stories the better.

It is rarely possible to use natural ventilation for music spaces because open windows allow a high degree of flanking sound transmission. Location of supply and return fans close to critical spaces served is rarely cost-effective because the savings in duct length are compensated by the added expense of silencers. Duct layouts for rows of practice and teaching rooms should place the supply and return truck runs over corridors with internally lined branches to each room as pointed out in Figure 9.15.

9.13.2 Privacy

Both speech and music privacy are signal-to-noise ratio problems. The intruding speech or music signal, which is the source-room signal reduced by the separating building construction, is compared by the listener with the background sound of the receiving room. The greater the intruding signal compared with the background noise, the less music privacy exists; correspondingly, the greater will be

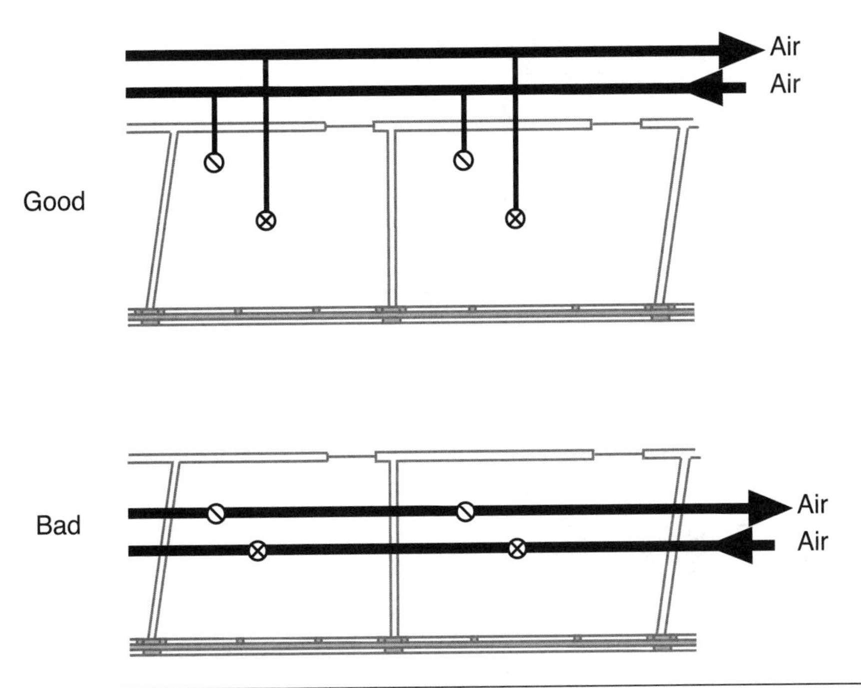

Figure 9.15 Schematic of effective and ineffective duct layout in music practice and teaching rooms.

the annoyance and interference effects of the intruding signal. Within a rather wide range of background sound levels, the typical listener does not differentiate between the reduction effect of the intruding signal and the masking effect of the background sound in judging the acceptability of the perceived signal-to-noise ratio. Again, this assumes that the noise and the speech or music signal have reasonably the same spectra.

The question arises: Is it possible to simplify the problem using single-number descriptors? For example, a system of criteria that uses laboratory-derived *STC* ratings or field-measured noise isolation class (*NIC*) ratings for sound-isolating constructions and commonly used A-weighted or C-weighted sound levels for background sound levels and for source music levels, would have widespread use in the early planning of music buildings even if detailed design should continue to octave-band analysis.

Table 9.1 represents one such single-number comprehensive approach to dealing with the full range of music teaching facility spaces. Allowable *NC* ratings and equivalent A-scale *SPL* values are given for background levels in each music space. The expected maximum source room levels are given in terms of 5 percentile (levels reached or exceeded only 5% of the time) C-weighted sound level values. The

Table 9.1 Acoustical privacy criteria for music buildings

	Maximum background level	Maximum background level	Band room	Orchestral room	Choral room	Organ practice	Music classroom	Music listening	Practice room	Ensemble room	Electronic music	Faculty studio	Assumed maximum level
	NC	dBA											dBC
Band room	25	36	65	65	65	62	65	65	57	61	62	61	103
Orchestral room	25	36		62	62	62	62	62	56	59	62	62	100
Choral room	25	36			59	62	59	59	56	59	62	62	97
Organ practice	34	44				54	62	62	54	58	58	58	100
Music classroom	25	26					59	59	56	59	62	56	97
Music listening	25	36						59	56	59	62	56	97
Practice room	35	44							48	52	54	52	94
Ensemble room	30	40								55	58	55	97
Electronic music	30	40									58	58	100
Faculty studio	30	40										52	94

NIC number is a single number representing the *STC* value discussed earlier and converted to overall sound isolation by the formula:

$$NIC = STC - 10\log(S/A_2)$$

where S is the area (in sq ft) of wall separating the two spaces and A_2 is the total receiving room absorption averaged over the octave bands 125 to 4000 Hz.

In using Table 9.1, one must remember that structure-borne isolation may be critical. Many musical instruments such as cellos and double basses, set up floor vibrations directly. Also, the table assumes the use of relatively heavy constructions with reasonable isolation at low frequencies. Applications of the table to some complex wood and drywall or multiple drywall constructions may be optimistic. A one-third octave band or octave-band analysis would show the deficiencies of these constructions at low frequencies. The past experience of teachers using the facilities, the need for multiple uses of a single facility—with the most critical use governing—and the overall budget limitations of the project must be considered when applying the table.

The most critical spaces, including concert, chamber, recital halls, studios, and most worship spaces, require separate analysis because their acoustical requirements always require special study. Background levels as low as *NC*-15, or 27 dBA, are desirable, and intruding sounds from adjacent spaces should be inaudible. *NIC* values of 75 or higher are required, and that means that these critical spaces should be surrounded with lobbies, corridors, storerooms, and similar buffer spaces. Speech privacy considerations are somewhat less critical and demanding. For lecture-only classroom isolation, for example confessionals in Catholic churches, 200 mm solid or sand-filled block or concrete or common brick is adequate. Lighter, equally effective but more complex drywall construction is possible (see Figure 9.11). Still, airtight construction and full gaskets on doors are required.

Engineering balance is a key goal. For example, partitions separating critical spaces from corridors do not need to be as effective as those separating the critical spaces. Similarly, the common partition construction must be matched to the floor and ceiling construction in order to achieve balanced noise reduction laterally as well as vertically and to minimize flanking sound transmission paths. Figures 9.13 and 9.14 show typical balanced construction for music spaces requiring performance levels in the range of *NIC* 65, both vertically and horizontally. This type of construction would be appropriate for music teaching and listening rooms, and some might find it applicable even for practice rooms if funds are available to permit a more than minimum standard. Low ceilings for band, orchestral, and choral rehearsal rooms can result in excessively loud performance levels and inadequate inter-musician communication. Ceiling heights of 5.5 meters or more are required for optimum results for larger performance and rehearsal spaces.

There are now pre-manufactured practice room and teaching modules that can be inserted into a large space and that provide adequate isolation and room interior treatment. They frequently include the use of electronic masking noise to represent the masking noise of ventilating systems. Prefabricated modules are particularly cost effective for adaptive reuse of an existing building for music instruction, and electronically generated masking noise is particularly desirable when the air supply system is a variable volume type.

9.14 SOUND-ISOLATING WINDOWS, PARTITIONS, AND DOORS

The constructions suggested for music teaching and practicing facilities also have applicability for critical sound isolation tasks such as recording, broadcast, and television studios. Figure 9.16 shows a typical construction for a window between a broadcast or recording studio and a control room.

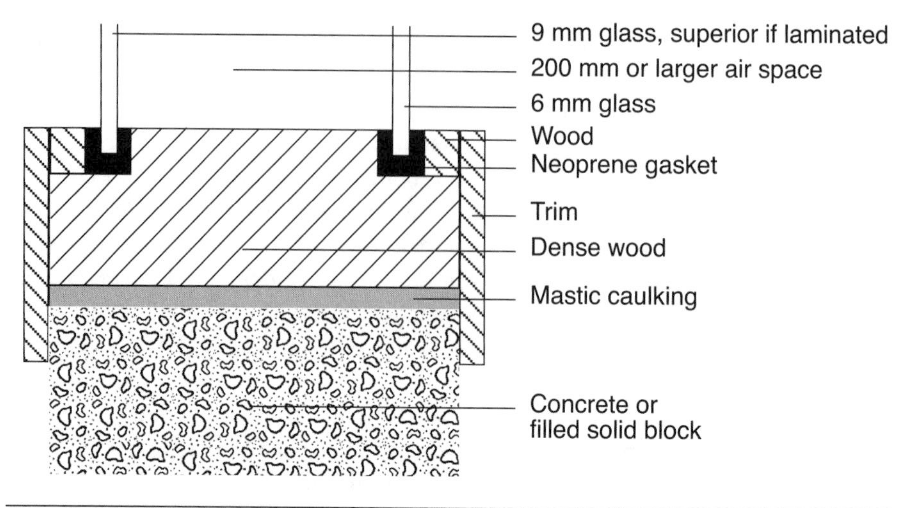

Figure 9.16 Design of a sound-isolating window.

Figure 9.17 A large, commercial metal partition. The doors turn to close (photo: IAC, Bronx, NY).

Some spaces are designed with movable metal partitions for flexibility. Commercially available sound-rated partitions such as seen in Figure 9.17 have sound-isolating value, especially if installations can be airtight at perimeters.

Many special sound-isolating doors are available, but at considerable expense. Good, workable gaskets that provide airtight closure every time are the most important factor for door sound isolation. Good gaskets can be applied to standard doors to improve their sound isolation. Hollow-metal doors, similar in construction to Figure 9.17, are better than solid-core wood doors, and solid-core wood doors are better than hollow-core wood doors. Louvered, poorly fitted, and undercut doors provide negligible isolation. Notable special cases are the door walls, which are the movable partitions typically used to separate a worship space from an adjacent social hall or auditorium. They are opened during services that attract crowds, such as on Jewish High Holy Days, and closed at other times. Such partitions must be designed for high sound isolation to provide for simultaneous use of the two spaces. Several manufacturers provide excellent complete systems for such movable walls, including tracks, gaskets, etc. However, it is of utmost importance to investigate and control the flanking sound-transmission paths such as over the ceiling, requiring an airtight partition from track to floor or roof slab above as well as supply and return ductwork, and usually requiring silencers.

Figure 9.18 shows how a hallway between two critical spaces can result in more economical and lighter, movable partitions. The plan shown is standard for

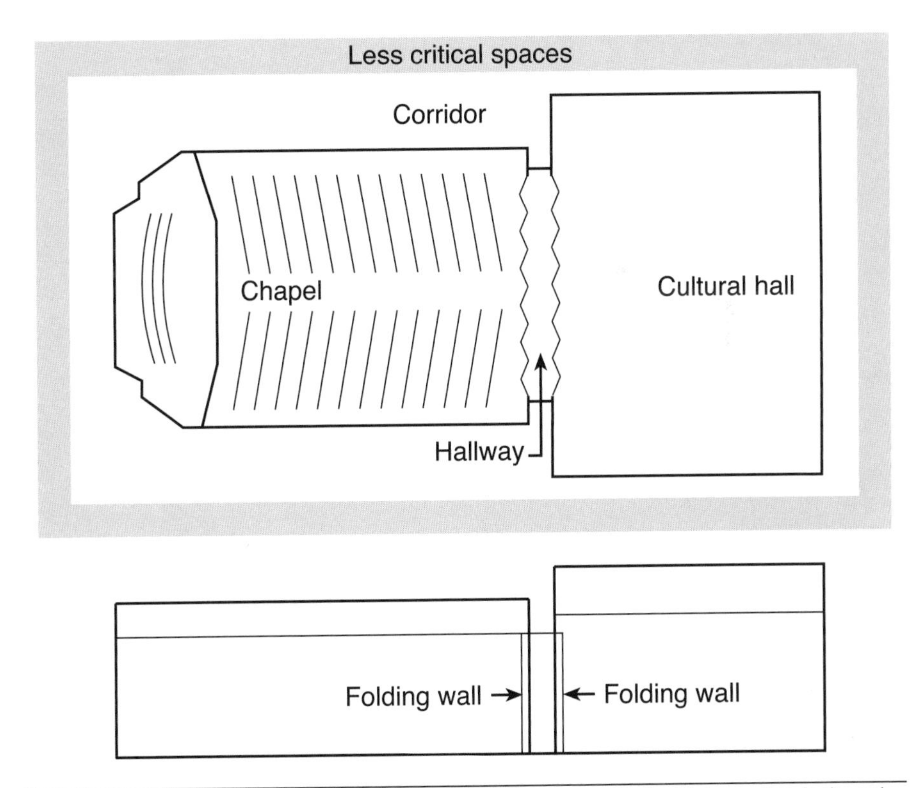

Figure 9.18 Typical Mormon Church Stake or Ward. The room with the higher ceiling is the cultural hall and the Chapel is the other room with a lower ceiling. Separating these two rooms is a sound-isolation hallway and two movable partitions that allow their use together or separately.

many Mormon facilities both inside and outside of the United States. Use of two areas with a hallway between allows simultaneous use without interference.

9.15 EXTERIOR NOISE

Practical examples, thus far, have treated sound isolation between spaces within one building, but the same principles apply regarding isolation from exterior noise. Ideally, a new building for worship should be located on a quiet site. Practically, one must deal with houses of worship located near major airports, highways with heavy truck traffic, railroads, factories, outdoor sports stadiums, and even auto racing tracks. The outside of the building envelope then comprises the source room with the critical space to be isolated as the receiving room, and the subsequent analysis and design follows the same rules as interior isolation. Surrounding critical worship and music rehearsal and teaching facilities with circulation spaces is an element of basic planning that should be adopted in such situations when possible.

10

SOUND SYSTEMS FOR CLARITY AND REVERBERATION

10.1 INTRODUCTION

10.1.1 Sound Level Amplification

Sound sources usually heard in auditoriums and worship spaces such as most musical instruments and the human voice, produce sound levels that are adequate for listening at relatively short distances in rooms of small volume. Most rooms in which sound is expected to travel more that 15 m (49 ft), and having a volume of 1500 m^3 (490,000 ft^3), require a sound reinforcement system for untrained talkers and certain solo musical instruments to ensure adequate sound distribution. Even in smaller meeting rooms, although strong-voiced speakers can be heard clearly, weaker voices can require reinforcement. For satisfactory intelligibility, voice levels should be 25 dB above mid-frequency background noise. Also, for certain operas and for many Broadway musicals, the voices of the singers and actors require amplification to avoid masking by the orchestra. Speech reinforcement systems generally require capabilities to ensure at least 90 dB undistorted audience peak sound pressure levels, classical music 105 dB, and music 125 dB SPL (see Figure 10.1) (see References 10.1 and 10.2).

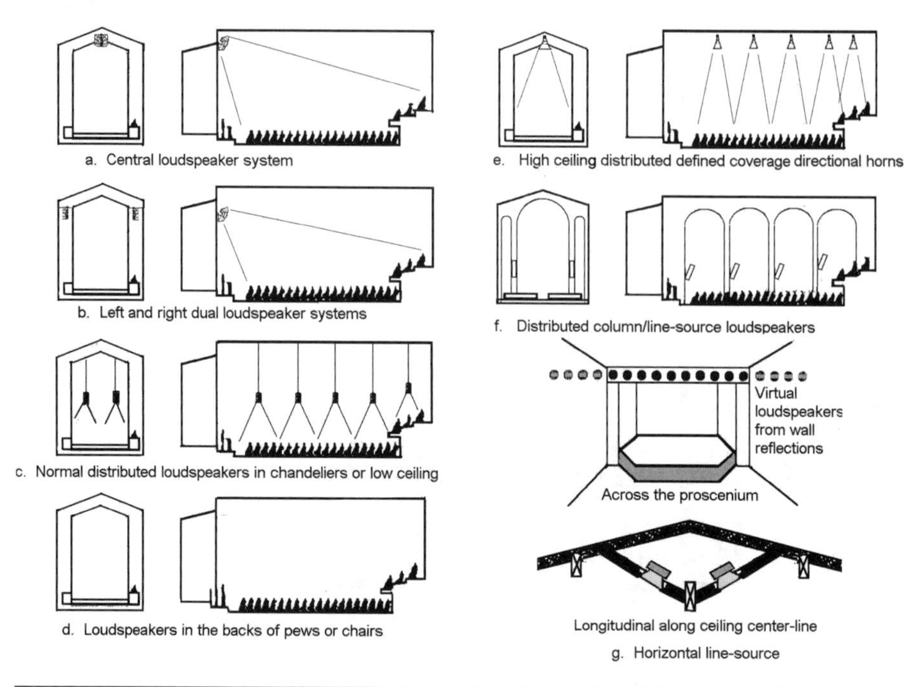

Figure 10.1 Sound systems designed for intelligibility: (a) Type I central system, (b) Type II split central system, (c) Type III conventional distributed cone system, (d) Type IV pew-back, seat-back distributed system, (e) Type V distributed directional horns, (f) Type VI distributed column loudspeakers, and (g) Type VII horizontal line sources across proscenium or along ceiling centerline (see References 10.10 and 10.11).

10.1.2 Increased Clarity or Increased Reverberation

In certain spaces, particularly cathedral-like spaces, one may want to use amplification to allow both high speech intelligibility and naturally long reverberation times, as well as warm, reverberant room acoustics for music in the same room. One can design a large, reverberant room with *cathedral acoustics*, then use a highly directionally controlled sound amplification to ensure clarity and minimization of sound energy directed into the hard-reflecting areas. Conversely, one can design a good speech-acoustics room and then use an electronic enhancement system to simulate a lively acoustical environment with a long reverberation time and reflections from many directions. While slightly over 2 sec is probably the upper limit for sufficient speech clarity through room acoustics alone, the existence of any upper limit for reverberation time that some fine sound system cannot tame for high intelligibility has not yet been proven (see References 10.3 and 10.4).

10.1.3 Frequency Response

Speech reinforcement systems generally require a bandwidth from 125 Hz to 10 kHz for complete naturalness although less bandwidth, 250 Hz to 8 kHz, can suffice for complete intelligibility. Music requires greater bandwidth—50 Hz to 16 kHz sufficing for chamber, choral, and most orchestral music; 25 Hz to 16 kHz for organ and all orchestral music. Intelligibility actually can be improved in many real room-acoustic situations by rolling-off sound energy below 500 Hz; where the reverberation time increases frequently, less directional control is performed by the loudspeaker system and the signal has minimal active contribution for intelligibility. The tradeoff between optimization for the two subjective evaluations of a speech amplification system, intelligibility vs. naturalness, can be adjusted for each situation (see Reference 10.2).

10.1.4 Freedom from Distortion and Noise

The frequency responses and sound pressure levels (*SPLs*) mentioned should be provided without distortion, rattles, buzzes, and other signs of overload. Distortions to avoid include harmonic, intermodulation, and slew-rate (rise- and/or fall-time for impulse responses). Wide, dynamic-range contemporary music is often the severest test for freedom from distortion.

10.1.5 Directional Realism

We would like sound to appear to come from the live sound source, even though most may come from a loudspeaker system located elsewhere. Proper loudspeaker locations for central systems and signal delays for other systems can optimize the use of the Haas or precedence effect. Slightly delaying the amplified sound a few milliseconds after the natural sound can suggest to the hearing mechanism that sound is coming from the natural sound direction even if the amplified sound energy is up to 10 dB louder (see References 10.1, 10.4, 10.6, 10.7, and 10.8).

10.1.6 Balance between Clarity and Spaciousness/Liveliness

In the chapters on worship spaces, we discuss that most worship music benefits from long reverberation times and a sense of omni-directional sound. These characteristics, which often come together in a reverberant building, can be provided by the introduction of multiple delays, digitally developed reverberation, and surround loudspeakers in electro-acoustic systems. They usually have a negative effect on speech intelligibility and music clarity and must be used cautiously. Thus, reverberation and liveliness systems are switched off during speeches in worship spaces.

Note that an articulation loss of consonants (*ALCons*) of approximately 15%, corresponding to an early-to-reverberant ratio of 4 to 5 dB, can be considered good for a lecture sound-reinforcement system, *ALCons* of 5 or less, with early-to-reverberant ratios of 10 dB or greater, are necessary for theatrical sound systems in which the originating speech can be a whisper or some other nonstandard professional speech delivery. In worship rooms, this condition results when lay, untrained speakers or children address older people with some hearing loss (see References 10.9 and 10.11). With many types of music, too much intelligibility representing clarity may be a negative factor, representing a lack of liveliness and reverberation (see References 10.3 and 10.4).

10.1.7 Sound System Uses Other than Reinforcement

Added loudness, clarity, or lack-of-clarity in terms of spaciousness and reverberation are only part of the story. Other possible sound system uses include:

a. Playback of recorded sound
b. Sound recording often simultaneous with reinforcement
c. Teleconferencing
d. Sound effects generation, recording, or playback for drama
e. Impaired-hearing assistance
f. Pickup for television and radio broadcasting
g. Latecomer lobby coverage
h. Program distribution and paging throughout a building

Many religious buildings have specific requirements that can be met by one or more of these systems.

10.2 BASIC TYPES OF WORSHIP SPACE SOUND REINFORCEMENT SYSTEMS

Systems that can provide intelligibility include:

a. Central
b. Split central
c. Conventional chandelier (low ceiling) distributed cone loudspeaker
d. Distributed pew or seat-back
e. Distributed directional horns
f. Distributed delayed column
g. Horizontal line source

10.2.1 Type I: Central Systems

The authors find the central systems most applicable for speech-reinforcement situations, with one or more loudspeakers in a single group providing reinforcement for an entire audience or congregation and for all microphone pickup locations. Figure 10.2 illustrates such a system with the loudspeaker at the front and center of the ceiling. Typical locations are above the talking or performing location, in the proscenium arch of the traditional theater, or in the chancel arch of a traditional church with directional characteristics and loudspeaker orientations intended to provide uniform coverage for the audience. Additionally, it serves for the reduction of apparent reverberation in reverberant spaces by minimizing sound radiation on wall and ceiling surfaces and restricting coverage to the audience area as much as possible. Many systems employ high-frequency horn loudspeakers and low-frequency boxed, direct-radiating loudspeakers in or just below the chancel arch of a traditional church with directional characteristics and loudspeaker orientations, again, intended to provide uniform coverage for the audience as well as a reduction of apparent reverberation in reverberant spaces by minimizing sound radiation on wall and ceiling surfaces and restricting coverage to the audience area as much as possible. The use of computer-steered-pattern array loudspeakers is more recent and may prove best for most applications in the future.

10.2.2 Type II: Split Central System

The split central system is frequently designed as a variant of the central system and applicable when there is not an architecturally satisfactory way of installing a central cluster and/or when much of the speech originates from the left and right sides of the front of the room. In a church, one loudspeaker system can be located

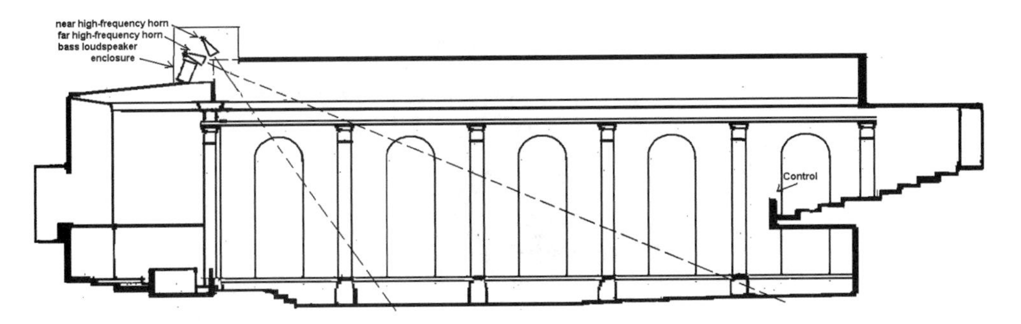

Figure 10.2 Type I central cluster, Immanuel Baptist Church, Paducah, KY (G. Anthony Johnson and Associates, Architects; Acoustics, Bolt Beranek and Newman). *Cross-eyed* design based on 1958 BBN system at Compton Laboratory Auditorium, MIT. This system is exposed (designed by Bolt Beranek and Newman) (see Reference 10.12).

over the pulpit on one side and another over the lectern on the other side, reducing the need to suspend a more visually apparent system from the ceiling centerline. In this case, both loudspeakers, individually, should provide complete audience coverage and only the pulpit loudspeaker should be used when the minister or priest is speaking from the pulpit. Certain situations can force the use of both loudspeaker systems at the same time. Reduction in clarity because of varying arrival times of the amplified sound, particularly in the front corner, should be evaluated for each situation. If left and right loudspeakers are less than 10 m (40 ft) apart, the problem usually does not exist, as shown in Figures 10.3a and 10.3b. The system shown in Figure S.15 has left and right upper- and lower-column loudspeakers. With careful selection of grille material and color, total invisibility is achieved.

For both Type I and Type II, the choice of directional horns or column loudspeakers usually is resolved in favor of directional horns, because they provide directional control in both horizontal and vertical planes, and column or vertical line-sources in the vertical plane only; but the newer programmable computer-controlled column loudspeaker systems are now state-of-the-art for many of these loudspeaker applications. A system for Harvard's Memorial Church, shown

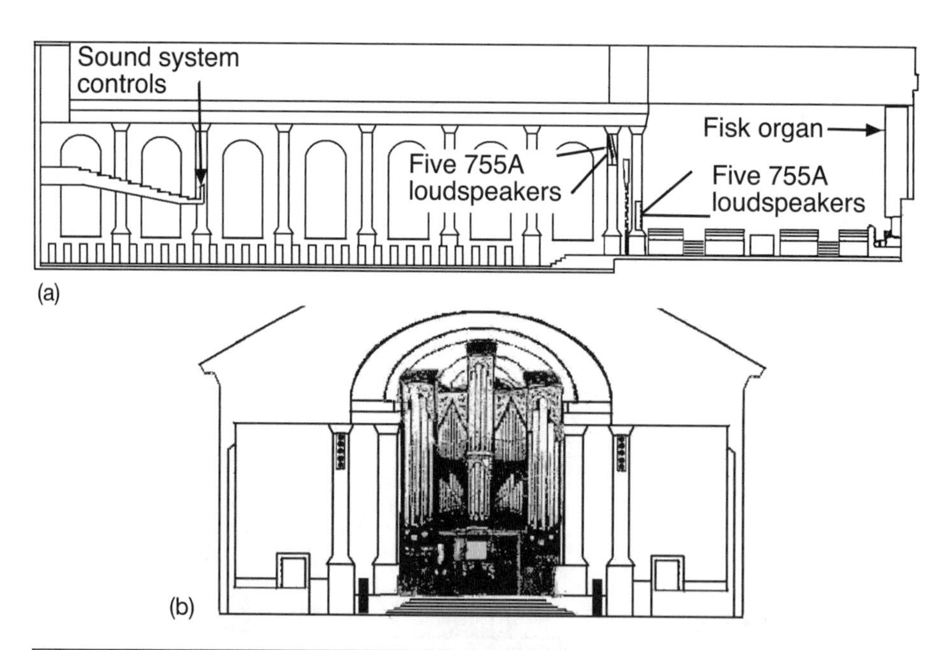

Figure 10.3 Longitudinal Section: (a) Harvard Memorial Church, Type II with basic column loudspeakers. Left and right systems are used separately and are now supplemented by under-balcony loudspeakers on digital delay. Original designer was David Lloyd Klepper at Bolt Beranek and Newman, and the system has more recently been modernized by Cavanaugh Tocci and Associates with original loudspeakers (see References 10.3 and 10.12). (b) Chancel elevation.

in Figure 10.3, employs two-column loudspeaker systems, one over the pulpit and the other over the lectern. This system has been in operation more than 40 years and is satisfactory, in part, because the space's low-room volume restricts reverberation time. Another system that has been in use for many years is located at the Immanuel Baptist Church in Paducah, Kentucky, where cross-firing two directional horns manages in a Type I central system in a minimal space situation (see Figure 10.2).

Note that it is possible to make columns of many directional horns to obtain a sort of *super column* for directional control in a reverberant environment. One of the best examples, in operation for over thirty years, is the Type II system at the Basilica de Notre Dame de Cap, Cap de la Madeleine, Quebec, Canada, providing high intelligibility and naturalness in a 4.5-sec fully occupied environment.

Generally, the higher the *RT*, the more controlled the central cluster must be and, consequently, the larger the radiating area for Types I and II. The large size of directional arrays can be overcome by careful architectural design or by the work of a sculptor, as with the slats hiding the loudspeakers in the Reform Synagogues Beth El and North Shore Congregation Israel (see Figures 10.4 and S.14).

Usually, central loudspeaker systems work best when some type of sound absorption is permanently present such as upholstered seats or pew cushions when full occupancy is not present. With hard seats, preferably, people should be grouped in defined areas and just those areas that are covered; this is sometimes

Figure 10.4a At North Shore Congregation Israel, Glencoe, IL, the cluster is located in the upper part of the Aron Ha-Kodesh. Minoru Yamasaki was architect for both synagogues. Acoustical Consultants: Beth El, Klepper Marshall King; NSCI, Bolt Beranek and Newman (see Reference 10.14) (photo: David L. Klepper).

Figure 10.4b Hiding Type I central loudspeaker system by sculpture. The central cluster high-frequency horns are well above the top of the ark, behind the splayed, dark wood slats, a sculpture by Lee du Sell, who also designed the Aron Ha-Kodesh at Congregation Beth El, Bloomfield Hills, MI (see Reference 10.13) (photo: David L. Klepper).

hard to arrange with central systems. There are, however, some systems that do work well with hard seats.

10.2.3 Type III: Conventional Distributed Systems

The most frequent application of the conventional distributed loudspeaker system is in low-ceilinged spaces using ceiling-mounted, cone-type loudspeakers (see Figure 10.5d), but high-ceilinged spaces can also use such systems (see Figures 10.5a-c). In practice, the square layout pattern shown in Figure 10.5c is easier to install than the hexagonal arrangement also shown. The conventional cone loudspeakers should not be much more than 5 m (17 ft) off of the floor, because the lack of directional control of the conventional cone loudspeakers frequently will result in too much reverberant (delayed) sound energy, reducing intelligibility, although much of that depends on the room factors. Higher ceilings also would demand more efficient loudspeakers if they were farther apart.

Figure 10.5a Chandelier System, Princeton University Chapel, Princeton, NJ, Altec 408 loudspeakers (see Reference 10.15) (photo: David L. Klepper).

Figure 10.5b Type III, conventional distributed systems: temporary chandelier system, Lavino Field House, Morristown Academy, Morristown, NJ, temporary system, Western Electric-designed Altec 755A loudspeakers. Bolt Beranek and Newman system for stockholders' meeting (photo: David L. Klepper).

Distance between loudspeakers and ears of listeners [m]	 Square pattern	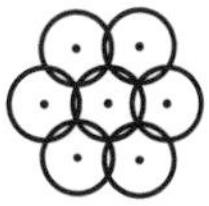 Hexagonal pattern
2	13,4	10,4
3	6,0	4,6
4	3,4	2,6
5	2,2	1,7
6	1,5	1,2
7	1,1	0,8
8	0,8	0,6

Figure 10.5c Type III, conventional distributed systems. Number of loudspeakers having a 90° distribution angle, per 10 m^2 for square and hexagonal layout patterns (see Reference 10.16).

Figure 10.5d Type III, conventional distributed systems. JBL low-profile cone-type loudspeaker control contractor 40 line for in-ceiling mounting (photo: courtesy of JBL/Harman International).

The low height of chandeliers and their density are the main problems for the use of these systems in high-ceilinged reverberant space. The chandeliers should be coordinated in conjunction with the seating plan; seating areas should be covered and hard aisles should not. Although visually the loudspeaker chandeliers may be designed to coordinate with lighting chandeliers, the design must not obstruct line-of-sight transmission from the loudspeakers to the listeners.

In a low-ceiling space with an acoustic tile treatment and a conventional distributed system, a distant listener only hears the nearby loudspeakers and not the direct live sound. In other situations, a combination of live and direct sound will be heard, and the use of digital signal delay to ensure coordination between amplified and live sound is essential for intelligibility, as well as useful for directional realism. Conventional distributed systems, on digital delay, are used to supplement Types I and II systems in under-balcony and balcony spaces in theaters, particularly for portions of the audience masked by balcony overhangs from the line-of-sight to the central systems.

10.2.4 Type IV: Pew-back System

A somewhat newer form of distributed system involves the use of a large number of loudspeakers in the backs of seats or pews. They can be operated at a low level and still be heard satisfactorily (see Figures 10.1d and 10.6. This system can be the equivalent to giving every listener a high-quality radio. The original loudspeakers used for this application were the KLH 6.5 and KLH 12.5 originally designed for that application. Do not skimp on the number of loudspeakers and do apply digital delay. Without a new development (discussed later), optimizing the delay for each loudspeaker and requiring each to have its own amplifier would not be economically feasible. Grouping loudspeakers into zones, each different by 15 ms from the preceding one, currently appears to be an effective practice. A 20 ms gap can be used, but more is not advisable. One loudspeaker for every two listeners is optimum, but one for every three is an acceptable compromise. The low level of

Figure 10.6a Example of pew-back loudspeaker enclosure installation: The Church of the Good Shepherd, Erie, PA (photo: David L. Klepper) (see Reference 10.17).

Figure 10.6b Example of pew-back loudspeaker enclosure installation: St. Thomas Episcopal Church, Fifth Avenue, New York City, NY (4″ loudspeakers used) (photo: David L. Klepper) (see Reference 10.18).

Figure 10.6c View of the nave and chancel of St. Thomas Episcopal Church, Fifth Avenue. This was the location of the first sound system to employ digital delay. Klepper Marshall King, consultants (photo from the church).

Figure 10.6d Continuous loudspeaker grille with pew-back loudspeakers (4″ loudspeakers used). Main floor, National Presbyterian Church, Washington, D.C., Harold Wagoner, architect (photo: David L. Klepper) (see Reference 10.12).

sound energy and proximity to a sound-absorbing audience negate the effects of high reverberation.

One unique example is the pew-back system designed by Peter Tappan for the 17th Church of Christ Scientist in Chicago. That system had to be distributed because of the possibility of speech pickup anywhere within the congregation, and the architecture ruled out a conventional or chandelier approach. Distributed microphones in the tops of chair backs complement the distributed loudspeakers in the back. A sophisticated automatic sensing system ensures that the loudspeakers switch off to avoid feedback in the area of the microphone that is switched on by the operator after seeing that a seat's microphone control button is lit by a congregant rising to address the remainder of the congregation and, thus, activating a seat-bottom switch (see Reference 10.20).

Because of the number of loudspeakers required and the desirability for signal delay and multiple amplifiers, seat-back systems can be expensive. This expense can be minimized with prepackaged miniature loudspeakers such as those manufactured by Tandy/Radio-Shack and used at Temple Rodef Shalom, Pittsburgh, Pennsylvania. They sacrifice a bit of visual elegance and further restrict low-frequency output, but still may be highly satisfactory.

10.2.5 Type V: Distributed Directional Horn System

A rare type of system that can be useful is illustrated schematically in Figures 10.1e and 10.7. Conventional cone-type loudspeakers would not work in high ceilings in reverberant spaces because too much long-delayed sound would be heard by lis-

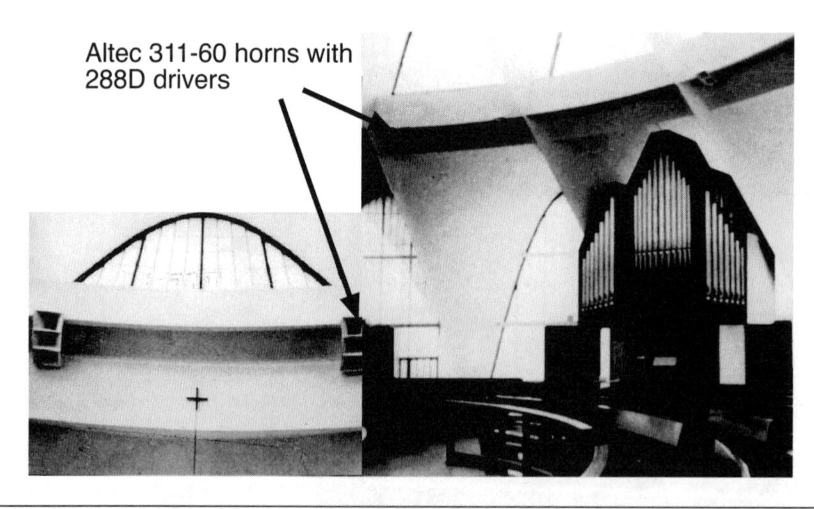

Figure 10.7 Horn loudspeakers in the St. Louis Priory Chapel. Type V distributed horns. Hellmuth Obata and Kassabaum, architects (photo: David L. Klepper) (see Reference 10.21).

teners. One can, however, design a horn-distributed system with each downward-pointing horn covering a defined area. Two examples are St. Norbert's Abbey, De Pere, Wisconsin and the St. Louis Priory Chapel, St. Louis, Missouri. The latter is illustrated in Figure 10.7.

The exposed horns obviously do little to modify the chapel's appearance. The geometry of the chapel and its completely hard and sound-reflecting interior pose problems for any sound system, and the empty reverberation time is in excess of 6 sec at mid frequencies. The vertically mounted horns are controlled by individual switches with the intent that only those loudspeakers covering people grouped together during periods of reduced occupancy will be used. The 30- by 60-degree directional characteristics of the horns effectively fit the individual areas to be covered and also keep sound energy off of the focusing circular walls. Still, consideration has been given to the replacement of this system by a pew-back system that requires less attention to loudspeaker switching.

10.2.6 Type VI: Distributed Delayed Column System

Distributed column loudspeakers using a delay were first applied by Parkin and Allen at St. Paul's Cathedral in London, England. Digital delay systems have replaced the more primitive methods used in that installation. See Figure 10.1f for a schematic design and Figure 10.8 that shows the design of column loudspeakers used at Duke University Chapel, Durham, North Carolina, that employs delays for loudspeakers to ensure directional realism. The individual line-source loudspeakers use the tapered line-source approach to assure constant directional characteristics over the speech range (see Figure 10.8). Figure 10.9 shows a *barber-pole*-style column loudspeaker and other methods to have frequency independent directivity.

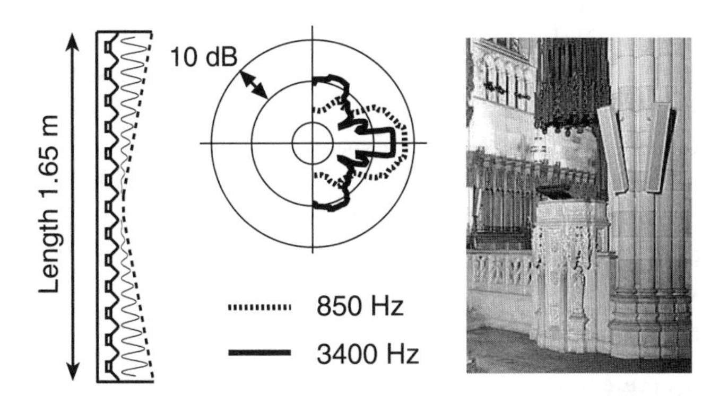

Figure 10.8 Past approaches to constant vertical coverage from a line source system. Duke University Chapel column loudspeaker. Designer of the sound system was Jacek Figwer at Bolt Beranek and Newman. Type VI System (see References 10.3, 10.4, and 10.22).

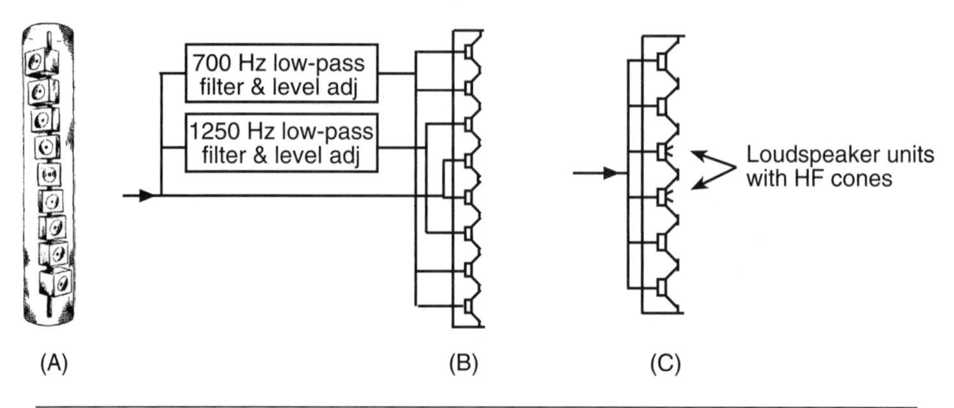

Figure 10.9 Past approaches to constant vertical coverage from a line source system: (a) Philips barber-pole line-source, Palais Chaillot, Paris. Type II system, (b) Old line source at Benjamin Franklin Hall, Franklin Institute, Philadelphia, PA, using electrical filtering. Type I line source, (c) University UCS-6 column loudspeaker system with only two center loudspeakers having extended high-frequency response.

The original glass-fiber loaded tapered line-source loudspeaker was part of a study project at the M.I.T. Acoustics Laboratory in 1956 and 1957 in collaboration with Henry Kloss at the KLH Company. The first application was as a Type I central cluster at Boston's Tremont Street Baptist Temple for which Doug Steele was the sound system designer. Intelligible and natural-sounding speech reinforcement was the only requirement. Today, in such a worship space, both much higher efficiency and sound output would be required or a second system would be provided for reinforced and recorded music and possibly other wide-dynamic range program material (see Reference 10.20). Figure 10.10 shows the distributed column system at Boston's Holy Cross Cathedral.

Today, there are techniques that preserve greater loudspeaker efficiency while shortening the line at higher frequencies as compared with glass fiber absorption in the path of loudspeaker output. Recent technology includes individual digital-signal processing control of each loudspeaker's individual amplifier with level, delay, and frequency response by computer-control setup. This type of digitally controlled loudspeaker is often referred to as a steerable-line (or steerable pattern) array source.

10.2.7 Type VII: Horizontal Line Source

More recently, there has been the use of the horizontal line source. Widely used by the Latter Day Saints (Mormon) Church, a horizontal line with 4.5-in loudspeakers alternating on either side of a continuous V enclosure along the centerline or peak of the church ceiling provides uniformity of coverage. This particular array system is schematically outlined in Figure 10.11.

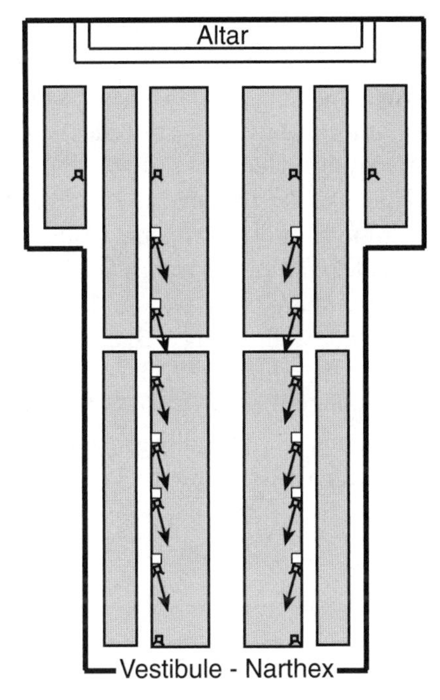

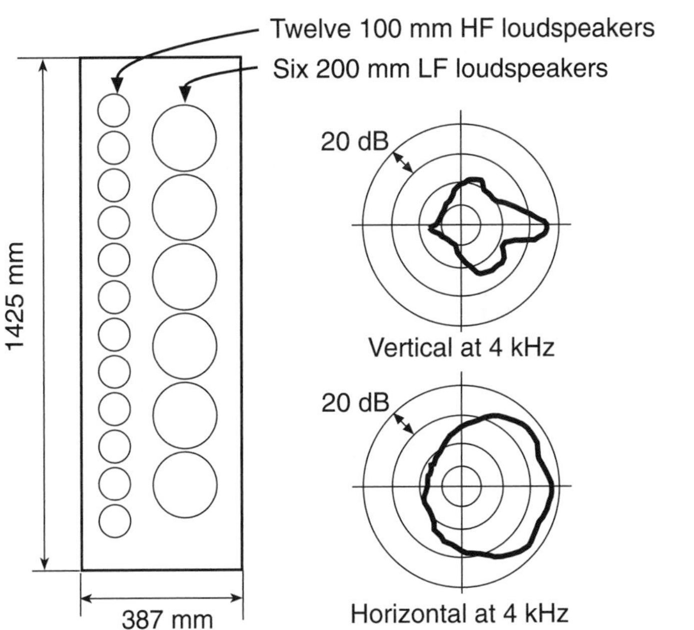

Figure 10.10 The distributed column system at Boston's Holy Cross Cathedral (photo from the cathedral) (see References 10.23, 10.24, and 10.25).

David L. Klepper and Larry S. King pioneered an *infinite* horizontal line across the top of the proscenium opening at the restoration and renovation of the Friedburg Concert Hall at the Peabody Conservatory, Johns Hopkins University, Baltimore, Maryland. The reflections from the side walls of the rectangular room effectively make the wall-to-wall line-source system an infinite line source. Similar systems are at the Marine Mammal Pavilion of the National Aquarium, also in Baltimore, and at Benjamin Franklin Hall, the American Philosophical Society, Philadelphia, Pennsylvania. See Figure 10.1(g) (see Reference 10.22).

There are other types of loudspeaker arrangements, including loudspeakers along side walls facing each other, and loudspeakers located without direct line-of-sight to the audience. With few exceptions, these are all poor choices for providing intelligibility but can be useful for music enhancement, including artificial room simulation with digitally generated reverberation and for simple music reproduction with a sense of mystery.

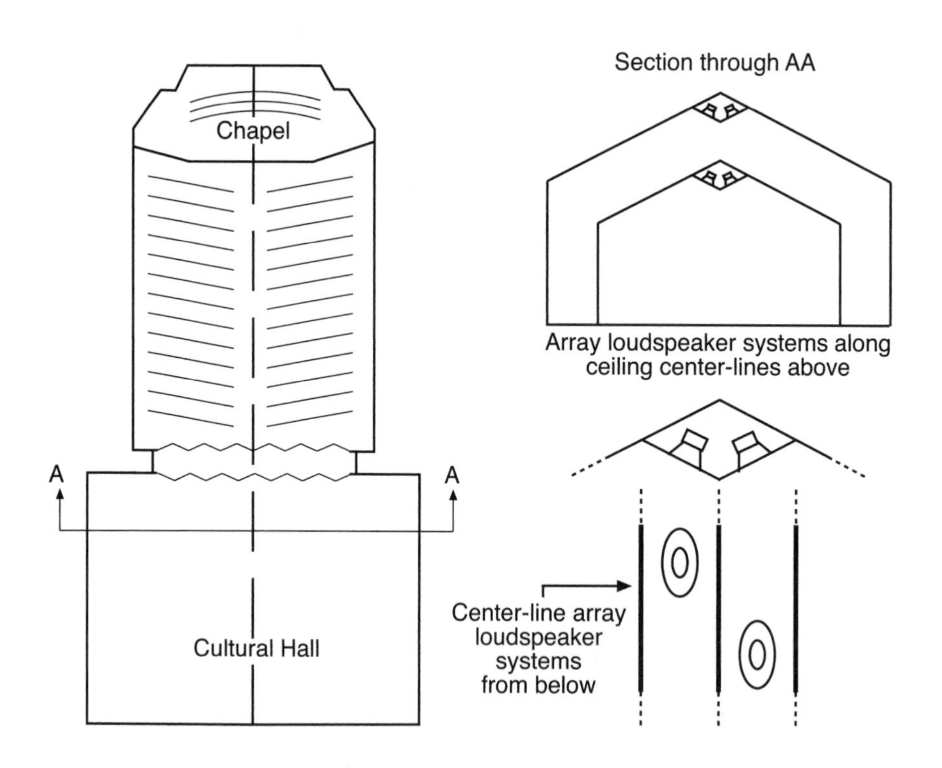

Figure 10.11 Typical Mormon Church stake or ward. The higher ceiling room is the Cultural Hall and the lower ceiling room is the Chapel. Horizontal array loudspeaker system along V-shaped ceiling center-lines in both Chapel and Cultural Hall (also see Figure 9.18).

10.3 EQUIPMENT

10.3.1 Microphones

There are four basic types of microphones: carbon, piezoelectric, dynamic, and condenser. The carbon and piezoelectric microphones rarely are used for sound-reinforcement work today, although much sound-system history is connected with their development and application. We will be concerned with the dynamic and condenser types. The dynamic microphones shown in Figure 10.12 are similar in operation to a generator. The movement of wire or ribbon conductors in a magnetic field, developed by permanent magnets integral with the microphone construction, causes the development of current and voltage. Ribbon microphones use a ribbon as the conductor and as the element moved by the fluctuating air pressure. Moving coil microphones cause the motion of a diaphragm responding to fluctuating air pressure to generate current in coils connected to the diaphragm.

In condenser microphones (see Figure 10.13), fluctuating air pressure causes one of two opposed plates of a capacitor or condenser to move toward and away

Figure 10.12 Representative dynamic microphones (not to same scale). Electro-voice Cobalt Co9 (left) Sennheiser MD431II_ZoomPro (right).

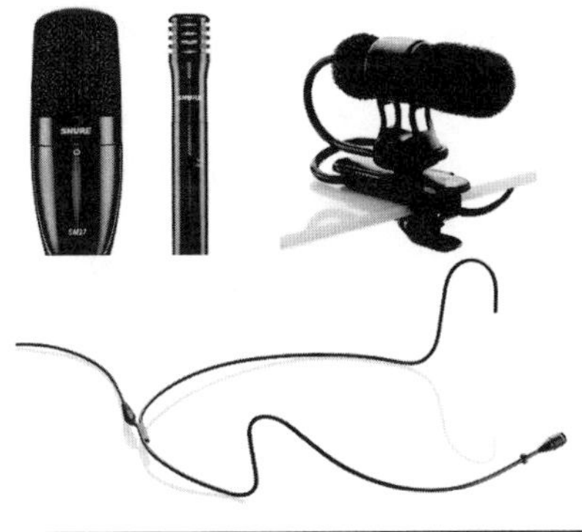

Figure 10.13 Representative condenser microphones (not to same scale). From left: Shure SM27 multi-purpose microphone; Shure SM137 multi-purpose microphone; DPA 4080 miniature lavalier cardioid microphone; DPA 4065 headband mounted microphone.

from the other. A fluctuating voltage is developed that can cause current. A DC polarizing voltage across the capacitor is needed. This can be provided electronically, but the *electret*-type condenser microphone includes permanent polarization.

Dynamic microphones are inherently more efficient and can drive microphone lines of reasonable length to a preamplifier that can raise the signal level of the relatively weak microphone current. The output of condenser microphones is usually weaker in terms of power. It is also high in impedance and, thus, subject to hum and radio-frequency pick-up and, as a result usually is immediately amplified by a preamplifier attached to the microphone before reaching a signal line. In the classic condenser microphones, the preamplifier base to which the actual microphone is attached also provides the polarization voltage. The power to the microphone preamplifier normally is furnished along the microphone line from a separate power supply or the control console or mixer-preamplifier, that is the control equipment, by a voltage between the shield and the two signal conductors. This is known as *phantom powering.*

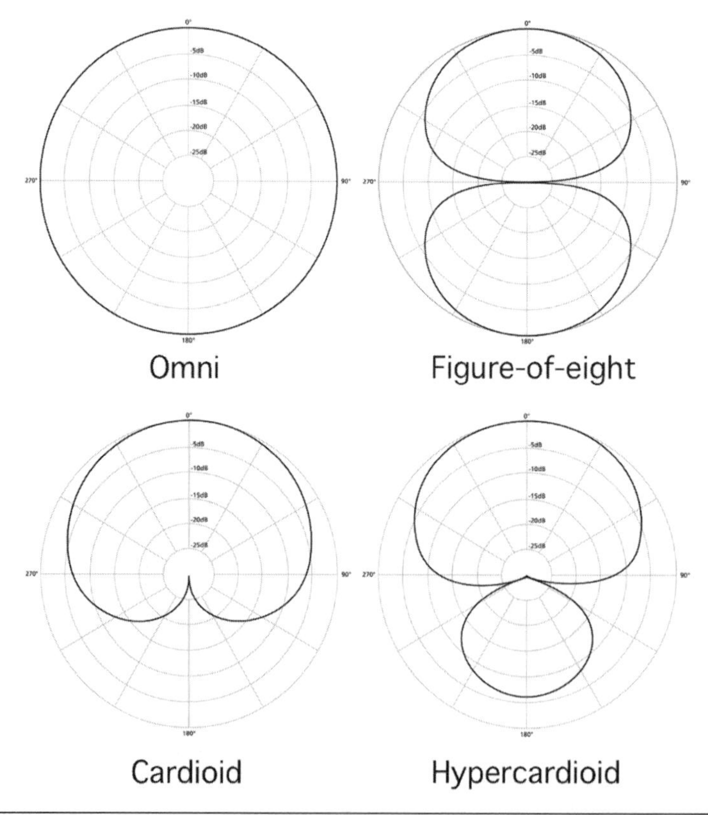

Figure 10.14 Polar plots of some common (theoretical) microphone directivity patterns.

Both dynamic and condenser microphones are available with several types of directional characteristics. Unaltered, the moving coil/dynamic and both types of condenser microphones are omnidirectional, that is, they pick up sound energy equally well from all directions. Actually, at high frequencies, where the wavelengths are similar to the diameter of the condenser plate or moving-coil diaphragm, or to the overall diameter of the case, the microphone does become directional. A real directional microphone, however, has multiple paths for air to reach the microphone condenser plate or diaphragm, often directing some of the pressure fluctuation to reach both sides. Mechanical filtering and delay is used to shape the directional characteristics, some of which are shown in Figure 10.14, and various types are ultradirectional, unidirectional, cardioid, hypercardioid, and figure 8. The last is the basic characteristic of the ribbon/dynamic microphone, but its directional characteristics can also be shaped. Multi-pattern microphones are available that employ moving coil and ribbon together with a variation in the ratio of the signal from the two elements determining the actual directional characteristics. Today, all dynamic and condenser microphone types mentioned are available with good response characteristics for professional work.

Wireless microphone systems are essentially a combination of a small, usually condenser-electret-omnidirectional microphone, a small FM radio transmitter, and the receiving system. They are especially useful for people who have to communicate without staying at one fixed position in the room. Figure 10.15 shows a typical wireless microphone system.

Ultra-directional microphones will be based on continuous or discrete arrays. One is displayed in Figure 10.16. The length of the microphone must be much larger than the wavelength so that the microphone has high directivity.

Figure 10.15 A wireless microphone system consists of a microphone, radio transmitter (possibly built into the microphone), and a radio receiver. Shown is the Audio-Technica wireless system AEW-4313 (from Audio-Technica Web site).

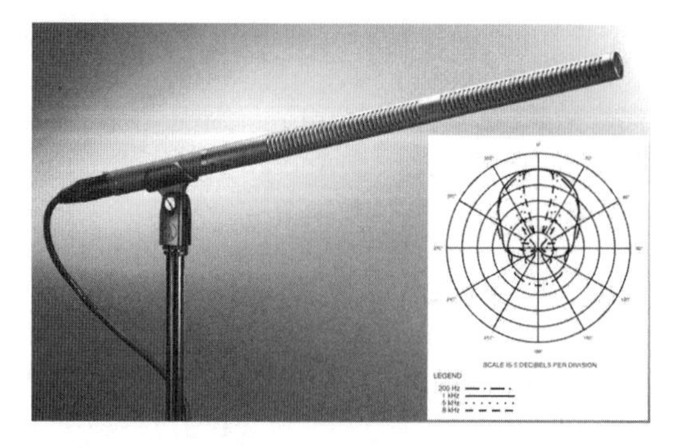

Figure 10.16 Ultra-directional microphone by Audio-Technica AT8015. Insert shows the polar pattern (from Audio-Technica Web site).

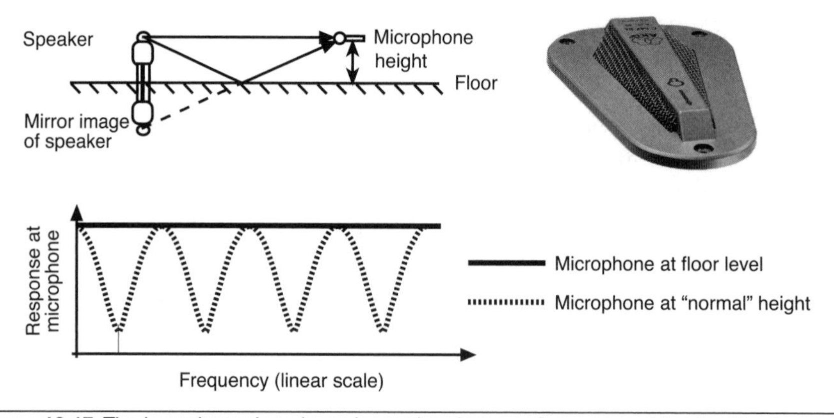

Figure 10.17 The boundary microphone is used on large surfaces to avoid coloration by reflected sound (see also Figures 3.5 and 7.3). Figure shows operating principle, frequency responses, and a commercial stereo boundary microphone, the AKG C547 stereo.

Finally, a boundary microphone shown in Figure 10.17 is intended to be used on large table, wall, and floor surfaces with characteristics designed for the modifications in directional frequency response that such applications demand.

10.3.2 Contact Pick-up Devices

Certain instruments used by popular music groups have their sound picked up by what are essentially vibration pickups. These usually operate on a magnetic (dynamic) principle, but piezoelectric and condenser elements have been employed. Most frequent applications are guitars (indeed many guitars are built specifically for such applications), banjos, violins, violas, cellos, basses, keyboard instruments, and harps.

10.3.3 Preamplification

There are preamplifiers in microphone bases, essential with condenser microphones, and preamplifiers that are part of the control equipment at the other end of the microphone line. The latter may also contain the necessary electrical connections for phantom powering. Today, analog to digital converters are included frequently at either location, and all subsequent signal transmission and processing takes place in the digital domain, including final power amplification and output to the loudspeakers.

10.3.4 General Control Equipment

If we include the preamplifiers, this equipment provides voltage gain, level control, frequency response, and input and output channel selection. In the simplest systems, these functions can be combined with the power amplifier into a single package *mixer-amplifier*. More complex systems have a control console, a control console with digital memory and presets, an automatic mixer-preamplifier, manual mixer-preamplifier, or any combination and number of these units.

10.3.5 Operated vs. Automatic Systems

During the history of sound reinforcement systems, certain systems were simply adjusted during installation and then left to be switched on and off, at times even simply left on. More complex systems, however, required hands-on control. This involved several functions: (1) choosing between different inputs; microphones not being used were turned off to prevent unwanted noise and reverberation pickup, (2) overall level control; keeping loud voices or musical instruments from blasting and ensuring soft ones were heard, (3) establishing balance between inputs, and (4) changing frequency balance when necessary. Even earlier systems had some form of limiters or compressors to avoid overload and/or to raise levels of low signals. The operating characteristics of compressors and limiters are indicated in Figure 10.18.

The first large systems that featured completely automatic operation were ones in transportation terminals such as airports. These systems introduced noise operated automatic level adjustment and was first applied at Dulles Airport in Washington, D.C. This device allowed levels in the terminal to be moderate during quiet periods and reasonably loud when crowd noise was present.

Today, various forms of *smart electronics* have given us automatic microphone mixer-preamplifiers and other control equipment that can accomplish many more functions, including:

a. Recognizing the microphone at which a person is talking, essentially raising gain for that microphone while suppressing others

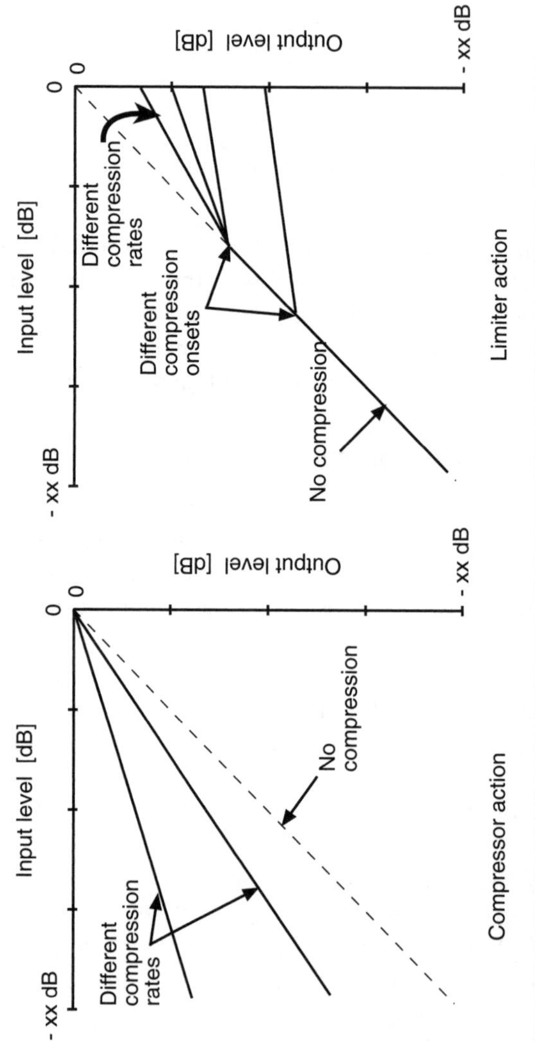

Figure 10.18 Output vs. input characteristics for a typical compressor (left) and limiter (right).

b. Reducing gain appropriately when people are talking to two or more microphones, reducing the chance for feedback
c. Keeping overall signal-level variations within limits to avoid system overload or blasting and to ensure intelligibility of weak voices, but maintaining the gain during periods without signal until the next signal can be analyzed
d. Providing priority muting if necessary
e. Providing the automatic routing of signals and logic control to switch to certain loudspeakers when people talk to certain microphones.

Oftentimes, such automatic systems are delegated to the routine tasks in complex theater-reinforcement systems so that the operator can perform only those tasks for which he is needed. In some systems, all these tasks are handled by digital equipment.

In certain cases, *geographical consoles* in which a map of the microphone layout includes buttons or other controls in a pattern that is a miniature of the actual location layout are an optimum solution. The 17th Church, Christ Scientist control, designed by Peter Tappan and Bob Ancha, is an excellent example (see Figure 10.19).

There are many operating *philosophies* for sound systems in worship spaces. A liturgical traditional church, Conservative synagogue or the typical mosque need not require an operator with normal services. A variety of automatic mixing and control equipment, both digital and analog, can be used to ensure intelligible and natural-sounding speech from a well-designed sound system. On the other end of the scale, there are evangelical churches featuring *Praise Band* music in which the sound crew is part of the creative team that assures both worshippers in the church and those viewing TV or listening to the radio that they receive the most relevant music and speech quality for their spiritual requirements. In some of these churches, there are three sound operators on separate, large and flexible control consoles: one for the worshippers in the church: one for radio, TV, and

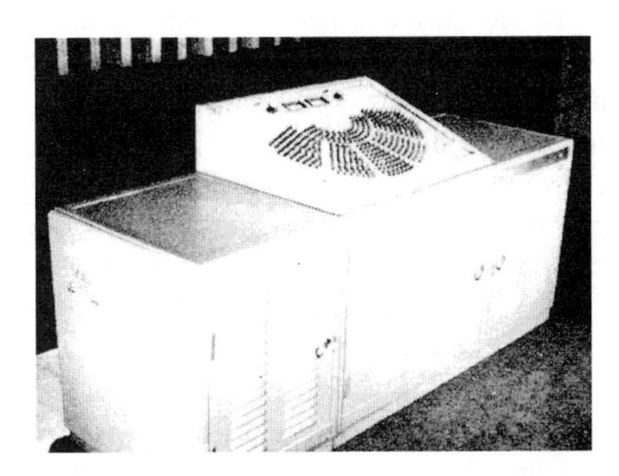

Figure 10.19 Geographical control console at 17th Church, Christ Scientist, Chicago, IL (see References 10.20).

Figure 10.20 Balcony control station and console at the Calvary Assembly of God, Orlando, FL. The first large church with a full delayed distributed system and a broad-front central system, used together or independently. Architects: Schweizer Associates. Acoustical Consultants: Klepper Marshall King Associates (photo: David L. Klepper).

sound to ancillary spaces: and one for the on-stage monitors and earphones used by the clergy and the musicians. In some cathedrals, normal events are conducted with an automatic microphone mixer, but a flexible control console is available for special events such as services or masses (see Figure 10.20).

10.3.6 Control Consoles and Mixer Preamplifiers

The simplest mixer-preamplifiers and control consoles have rotary knobs similar to home equipment. The more complex and professional consoles, typical of theaters and large *electronic* worship spaces, employ straight-line attenuators for all gain controls that may be operated frequently. Rotary knobs are reserved for equalization controls, *panning* controls that place the output of a given input to varying degrees in separate output or *group* mixing channels (for stereo effects, for example), and for trimming controls that may ensure that unusually loud microphone signals do not overload sensitive microphone preamplifier inputs. State-of-the-art control consoles convert the analog signal into digital immediately after preamplification and all signal processing (gain, equalization, and panning) is done in the digital mode. Both digital outputs, to feed digital recorders, and analog, to feed sound reinforcement, effects, playback, and monitoring systems, are available. These consoles usually include memory presets.

As noted, equalizers can be built into control consoles, mixer-preamplifiers, and mixer-amplifiers. When they are built into consoles, their purpose is most often to effect a change in the timbre or tone quality of the amplified sound. However, separate equalizers oftentimes are intended to compensate for room and loudspeaker characteristics and/or to ensure the greatest available gain before feedback. Frequently, equalizers built into mixer-preamplifiers and mixer-amplifiers are used for all three functions at the same time.

10.3.7 Controls Usually Fixed

The most frequently encountered equalizers in professional sound systems are 1/3-octave-band, octave-band, parametric, and notch equalizers. The 1/3-octave-band equalizers cut and boost, or cut only, in adjacent 1/3-octave-bands with some inevitable (and for some situations desirable) interaction between bands. Typical total variation for a cut and boost equalizer is +12 dB. The affected 1/3-octave-bands have center-frequencies from 25 Hz to 16 or 20 kHz. Often band-limiting and general treble-boost circuits are included in these devices as well. The octave-band equalizers are similar, but with one-third the number of bands. Our experience has been that the 1/3-octave-band equalizers are the most useful and most easily adjusted for a wide variety of sound system types.

The parametric equalizers may have the same boost-cut range, or may be cut-only, with gain reduction as much as 50 to 60 dB. The sharpness of the equalization curves and the equalization center-frequencies (position of maximum cut or boost) are also controllable. Often, the equipment has three, four, or five separate equalization circuits available, so these equalizers are the most powerful to deal with particular room and loudspeaker problems and are preferred by several sound system contractors and consultants.

Notch filters are intended exclusively for feedback control. They are really a simpler form of the parametric equalizer with four, five, or six circuits (and resulting notches in the frequency response) available in one piece of equipment. The attenuation curves are always sharp, allowing a narrow band around each single frequency that causes feedback or *ringing* to be attenuated sharply.

Normally, equalizers are designed to operate at line level with +12 to +20 dBm input and output levels and 150 to 600 ohms impedance. Distortion and *flat position* frequency response are similar to control and preamplification equipment. Today, this equipment is combined in packages with delay, level control, crossover, and signal routing equipment, all controlled by a computer connection that may be removed after final system adjustments. Sometimes, this equipment may be switched between preset memory positions.

10.3.8 Location of Controls

Generally, although automatic mixer-preamplifiers and other control equipment are rack-mounted with other electronic equipment, control consoles designed to be operated are located in desks placed where the operator hears the mixture of amplified and live sound in a manner similar to the audience. Complex entertainment and worship-space systems (for the electronic church, for example) may divide the control responsibilities among two of three operators with separate control consoles possibly fed by the same microphone. This is done for audience sound reinforcement and foldback or provision of an amplified sound back to the talkers and musicians (handled carefully to avoid feedback, but common practice today) and for recording and broadcasting.

10.3.9 Delay Equipment

Sound energy travels through air at approximately 344 m/s (1130 ft/s), varying to a slight extent with temperature and humidity. Some sound amplification systems can benefit from the application of signal-delay equipment to synchronize the sound from nearby loudspeakers with the sound from distant loudspeakers and/or with live sound from a distant speech or music source. Typical applications include delayed under-balcony, church and synagogue pew-back, and distributed-column loudspeaker systems. Typical delay times for these applications can vary from 15 ms out to nearly 1 sec for mega-church systems. Most systems that employ signal delay require multiple delays—two, three, and four delays are very common.

Another use of signal delay, involving short delays from 1/20 ms to 1 ms, is synchronization of loudspeakers in close proximity in a multiple-loudspeaker cluster. This can assist in avoiding any rough frequency response, resulting from phase cancellation in areas covered by more than one loudspeaker in the cluster in certain cases for specific areas.

The earliest signal-delay equipment employed tubes with a loudspeaker driver at one end, microphones along the length of the tube, and a sound-absorbing termination at the far end. Then, continuous tape-loop and magnetic disk record-playback-erase devices became available. Current practice is the use of digital delay devices that convert the analog signal to digital, store the digital signal, retrieve the stored signal for playback at various times, and then convert the signal back to analog. Sixteen-bit converters were once typical, giving somewhat over 90 dB dynamic range, but now 24-bit products are available with the lack of added coloration appropriate for the finest music systems. Often delay equipment is combined with other signal-processing devices in one digital microprocessor that can be programmed in a wide variety of configurations, including all-digital systems from microphone input to amplifier output. These can include delays, crossovers, equalization, automatic level control, and even input selection.

10.3.10 Feedback Protection

Although most readers, like the authors, have been exposed to sound systems that have *howled* because of too much loudspeaker energy reaching the microphone, calculations can be performed to ensure that a system design has plenty of potential acoustic gain (*PAG*) to meet the system requirements known as the needed acoustic gain (*NAG*). Factors included in this calculation include the distance of the talkers or other sound sources from the microphones, the number of microphones open at one time, the expected sound levels of the sound source, the directional characteristics of the microphones (a relatively minor role, often measurement microphones can be assumed to be omnidirectional, especially in rooms with more than modest reverberation), the number of microphones open at one time, the needed sound level at the most-distant or least-covered listener location, and the directional characteristics and intended point of the loudspeakers. All the *PAG* and *NAG* calculations assume equipment of good quality with reasonably smooth frequency response and uniform coverage characteristics (see Reference 10.26). Given a good design, with a designed-in safety factor, feedback or howl should not be a problem. However, sound systems sometimes are used in ways their designers did not intend. For example, a system designed for stage or platform or pulpit microphone pickup may have some comedian drag a cord microphone or use a wireless directly under a delayed under-balcony loudspeaker at the rear of the hall. Also, various factors may result in a compromised sound system design in which a normal 5 dB safety factor cannot be achieved.

Fortunately, beginning with Manfred Schroeder's Bell Laboratories-developed frequency shifter in1960, there have been a variety of economical electronic solutions to this problem. If the problem is not severe and occurs only at one frequency, then a simple approach is to return to the equalizer and simply reduce gain on the appropriate frequency band slightly, and then to check that both the problem is solved and that any changes in perceived sound quality are either not noticeable—neither degrading, unpleasant, nor resulting in reduced intelligibility.

There is a trend to equipping new systems with hardware that can cure more severe feedback problems. The electronic control package that performs many functions, including analog to digital conversion, control of the number and selection of open microphones, equalization, limiting and compression, and crossover signal routing, often has a feedback protection function that uses notch filtering that inserts a dip or notch in the frequency response precisely at the feedback frequency. Some equipment is automatic and continually analyzes the signal and inserts the frequency response notch at the precise frequency and to the needed depth. Frequently, this is combined with a degree of frequency shifting. The early Schroeder frequency shifter shifted the frequency a few Hz, enough to break the feedback loop but not enough to be noticeable in speech. But it was noticeable on nearly all types of music, and the effect was disagreeable. Modern frequency shifters continually modulate the shift about zero frequency change, moving from +1.5 Hz to −1.5 Hz, and the effect on music is not unpleasant, merely a slight vibrato, and then the feedback loop is broken.

10.3.11 Crossovers

Like equalizers, crossovers modify the frequency response (see Figure 10.21). The simplest variety directs the high-frequency signals, above 500 to 800 Hz, to the high-frequency loudspeaker or loudspeakers, and the low-frequency signals, below 500 to 800 Hz, to the low-frequency loudspeakers. Three-way and four-way

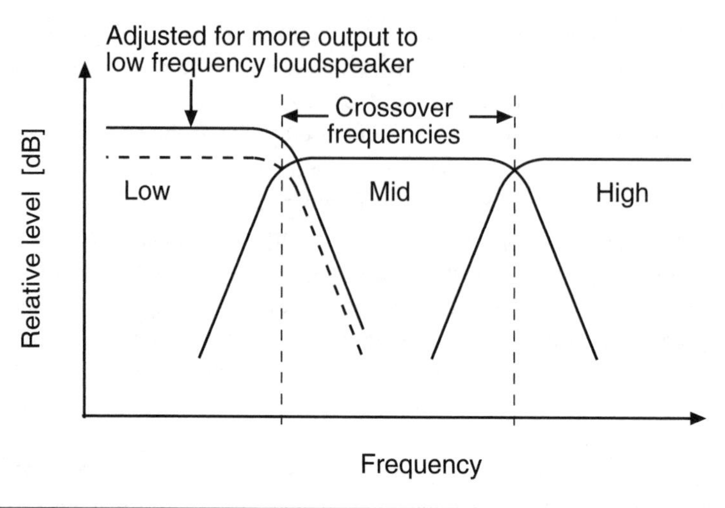

Figure 10.21 Crossover action sends appropriately filtered sound signals to the loudspeakers.

crossover devices are possible, directing signals to subwoofers, woofers, mid-range loudspeakers, and high-frequency loudspeakers (tweeters). There are several excellent loudspeakers that operate best over a restricted frequency range; therefore, crossovers are desirable for many types of full-range systems.

Crossovers once functioned only between the amplifier output and the loudspeakers and, thus, had to handle sizable amounts of current. They were composed of capacitor-inductor frequency dividing networks and, somewhat less frequently, capacitor-resistor networks. The disadvantages of high-level crossover networks included the reduction of the *damping factor* applied to the loudspeakers by the amplifier in the frequency range near the crossover frequency, some inefficiency because of resistive losses in inductors and any resistors, and a restriction to the simplest types of networks because of the substantial cost. Lower-cost home music systems generally use this approach for frequency response division, and the typical 6″ or 8″ coaxial loudspeaker for a distributed sound system may simply have a capacitor in series with the smaller high-frequency cone.

Most sound reinforcement systems use low-level crossovers, preceding the power amplifiers, with separate amplifiers for loudspeakers covering different frequency ranges. Advantages include more precision in the shape of the filter curves, retention of high damping of the loudspeakers by the amplifiers, retention of efficient coupling between the amplifiers and the loudspeakers, flexibility in balancing amplifier outputs to account for differences in efficiency between loudspeakers covering various frequency ranges, the possibility of including signal delay and/or phase-shift for signal alignment (included in some electronic crossovers), and lower overall distortion due to the reduced interaction between signals of different frequencies in the power amplifiers.

10.3.12 Power Amplification

Performance Characteristics

Power amplifiers are the sound reinforcement and reproduction equipment having the fewest problems and the best performance. Frequency response capabilities beyond those required for any reasonable system (10 to 100,000 Hz) are possible, and distortion can be virtually unmeasurable and below 0.25% in nearly any commercial power amplifier. Amplifiers are available in power-output ratings from 10 to 7000 watts with the 100- to 500-watt output ranges being most useful. Outputs are low-impedance, generally 2 to 8 ohms, 70-volt, meaning that the output voltage is 70.7 volts at full-rated output, or 100 volt, the latter used particularly for European equipment. The 100-volt and 70-volt amplifiers are typically used for distributed loudspeaker systems with four or more loudspeakers (up to hundreds, in some cases). Twenty-five-volt distribution is also used, mainly in school multi-classroom paging systems.

Configurations

In addition to the choice among low-impedance, 100-, 70-, and 25-volt outputs, the system design can choose between single-channel amplifiers, two-channel amplifiers, multi-channel amplifiers (four or six channels, usually), and modular units that can provide a multiplicity of channels working from the same power supply and/or building blocks to construct channels of higher power output than the individual modules. The modular units are typically compact and useful in reducing the size of equipment rooms in large installations. For the lowest-cost, least-flexible systems, mixer-amplifiers are available that combine preamplifier(s), controls, and the power amplifier in one unit.

Heat

Ventilation and cooling are important. Many power amplifiers radiate heat power that is greater than their power output rating. This factor must be considered in the design of the buildings in which they are located. New types of power amplifiers, employing switching power supplies and other advanced techniques, can reduce the radiated heat.

10.3.13 Loudspeakers

Basic Types

Two basic types of loudspeakers are the most widely used in public address systems and sound-reinforcement systems: (1) electrodynamic cone-type direct-radiator loudspeakers and (2) horn loudspeakers with compression drivers. Other types of loudspeakers used for some applications include panel loudspeakers, full-range electrostatics, magnetic-plane loudspeakers, and other exotic types. Except for mentioning that the panel loudspeakers have certain specific applications for wide-dispersion from wall and ceiling installations, despite somewhat irregular frequency response characteristics and polar plots (irregularities often not audible in particular applications), we will restrict our discussion to dynamic direct radiator and horn loudspeakers with compression drivers, plus one exotic and revolutionary type of system that does not employ electronics.

Direct Radiator Loudspeakers

Direct radiator loudspeakers usually employ some type of paper cone, although beryllium, aluminum, plastic, and even diamond have been used. Diameters range from 2 inches up to 18 inches with larger types having both somewhat greater efficiency and far better bass response. The bass response also, to an important extent, depends on the size of the air volume behind the loudspeaker with more volume yielding a more extended bass response. Bass loudspeaker enclosures can include ports and ducts tuned to complement the loudspeaker and enclosure characteristics, extending low-frequency response beyond that of a sealed enclosure, but with steeper roll-off below the flat response region. Whereas the larger loudspeakers provide better bass response, their larger cones become increasingly directional at higher frequencies, assuming that the heavier cones are able to produce high-frequency signals. Coaxial loudspeakers with 200 mm and 300 mm diameters, with smaller high-frequency cones and built-in crossovers, are used widely in distributed loudspeaker systems with the small treble cones ensuring more even distribution at high frequencies. Many successful distributed systems exist with simple, 100 mm or 125 mm diameter cone loudspeakers. A typical 125 mm diameter loudspeaker has a frequency range from 160 Hz to 8 kHz with 90-degree coverage from 6dB-down-points up to 6.3 kHz, good for distributed ceiling (Figure 10.5) and pew-back (Figure 10.6) systems, and in multi-loudspeaker line sources and arrays. Figure 10.22 shows a typical direct-radiator cone loudspeaker with an internal concentric high-frequency unit.

A typical 15 in low-frequency loudspeakers would be used between 25 and 800 Hz and provide 60-degree coverage at 800 Hz, wider below. Efficiencies of direct-radiator loudspeakers range from 0.25% to 3%, depending on many factors. Use of several low-frequency cone loudspeakers in one large horn enclosure with mouth dimensions somewhat comparable to the wavelengths of the frequencies radiated and supplemented by bass-reflex loading can raise the efficiency to nearly 10%. A further increased level can be achieved by providing a more directional sound source. This practice was typical for motion picture systems in large theaters for many years, low-frequency systems complemented the efficient horn high-frequency systems used

Figure 10.22 Typical direct-radiator cone loudspeaker with an internal concentric high-frequency unit (photo: Mendel Kleiner).

and made the best of the available amplifier power. Because of their huge size and the difficulty of integrating them into the architecture of a worship space, large horn low-frequency systems seldom are designed into such systems. Instead, the efficient horn high-frequency loudspeakers used with direct-radiator, low-frequency loudspeakers in compact tuned enclosures with low-level crossovers are used, and differences in amplifier power compensate for the variance in efficiencies.

Figure 10.23 shows the different distortion characteristics of loudspeakers and amplifiers. Whereas amplifiers have a drastic increase in distortion once levels become large, loudspeakers have a much more gradual increase in distortion with power. If loudspeaker input power can be kept low, the system will have low distortion. This can be achieved by using horns (that are very efficient) or by using array loudspeakers (where the power is shared between many loudspeakers).

There has been a trend away from separate horn clusters and bass boxes toward arrays of boxes with integral horns and direct radiators. While this trend poses no

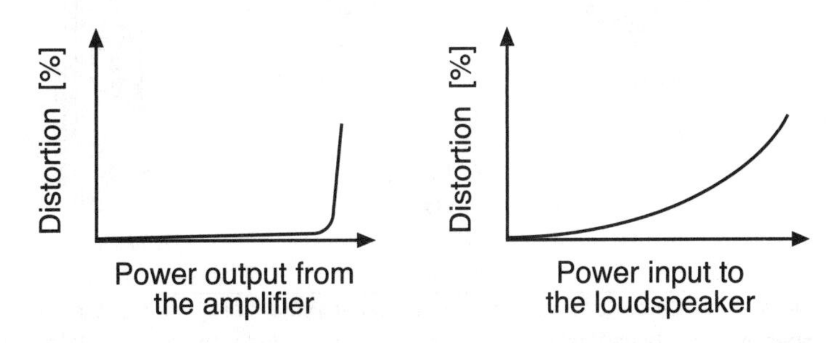

Figure 10.23 Distortion in typical loudspeakers and amplifiers compared.

problem for popular music reinforcement, it does reduce the effectiveness of engineered central clusters in maximizing speech intelligibility in reverberant spaces. But possibly the best approach in accomplishing such intelligibility enhancement today, for many situations, are the arrays or lines (columns) of many small direct-radiator loudspeakers with computer control of the input to amplifiers assigned one to one loudspeaker, controlling delay, frequency response, and a signal level to each. Intellivox (Duran Audio) and Eastern Acoustics Works pioneered this development. Examples of such systems are shown in Figure 10.24.

One advantage of the technology is that multiple beams can be implemented from one array, reaching, for example, both under and on top of a balcony. The beam can be adjusted to have different frequency response characteristics and, since the radiation angle can be steered away from the normal, they can be inconspicuously mounted flush against walls and pillars (or even inserted) where the conventional linear column loudspeaker would need to be angled away from the surface since it radiates to the normal of the column.

Directional Horns

Many horn sizes and shapes have been used with loudspeaker driver units. The purpose of the earliest horns was to improve the efficiency of the earliest loudspeaker drivers by providing more efficient coupling of the energy to the atmosphere. A side benefit was the increased directivity (where wide coverage was not desired or could be accomplished by additional horns) that further increases sound levels. Early horns were conical in shape and driven by sev-

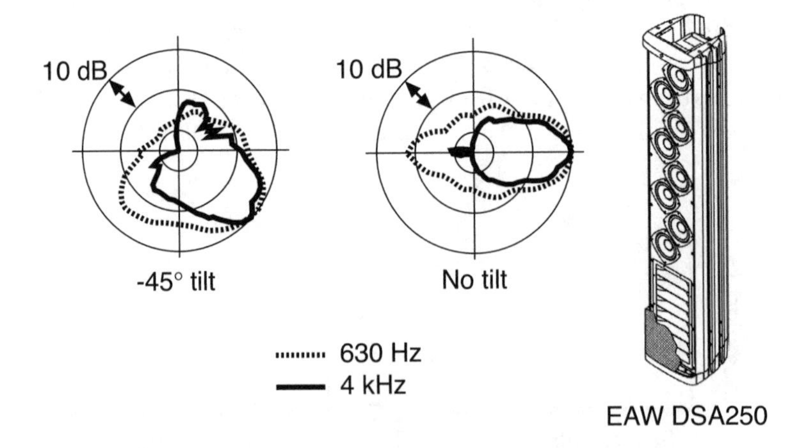

Figure 10.24 A steerable pattern line-array loudspeaker system and a few of the nearly infinite number of possible vertical directional characteristics (adapted from EAW technical information).

eral types of primitive driver units, including ordinary cone loudspeakers. The research that went into sound motion pictures resulted in great advances in the state-of-the-art at Western Electric and at RCA. Two basic types of high-quality directional horns evolved: (1) multicellular, with 2, 6, 8, 10, 15, or 18 individual exponential-expansion horns, forming cells connected to a common throat that preserved the exponential expansion from the compression driver's mouth and (2) radial, with exponential area expansion accommodated by curved upper and lower surfaces but with flat sides for controlled horizontal dispersion. The typical frequency response of such units was 0.3 to 8 kHz, and efficiencies were as high as 20% to 25%. Typical nominal coverage patterns ranged from 20 degrees by 40 degrees to 60 degrees by 110 degrees. These horn and driver designs have evolved into *bi-radial* or *constant directivity* horns. These have some flat and some curved surfaces in both planes. The exponential expansion is compromised to ensure maximum uniformity over the coverage pattern. A great variety of coverage patterns and frequency response ranges are available, and it is possible to design three- and four-way systems with cone low-frequency loudspeakers that easily cover the range 25 Hz to 18 kHz.

Recently, Altec, JBL, and E-V have developed skewed-coverage horns that provide higher levels and a narrower, horizontal pattern at the top of the vertical pattern and a wider, horizontal pattern and lower levels at the bottom of the vertical pattern. The horn loudspeakers can match many rooms' requirements, reducing the need for multiple-horns, single-loudspeaker systems (See Figure 10.25).

Figure 10.25 Almost invisible application of horn loudspeaker in the ceiling of Spring Hill United Methodist Church, Mobile, AL (photo: David L. Klepper).

Air Amplification

A recently developed microphone-loudspeaker system, initiated by Defense Research Technologies of Rockville, MD, is considered innovative in that it works without any mechanical moving parts or the use of electricity, except to charge a compressed-air tank or operate an air-compressor, blower, or fan. It uses the science of fluidics in a manner similar to fluid-control systems that regulate various processes in locations where the use of electricity is impractical. This approach uses a low-power air-stream to modulate the direction of a continuous, higher-power air-stream, and this motion may be detected as a fluctuation at higher pressure and velocity, thus producing sound amplification within the air domain without the conversion of sound energy to mechanical energy and then to electrical energy and back to mechanical energy and again to sound energy (see Figure 10.26). This technique is in its infancy but holds great promise, especially if it receives the kind of developmental effort that electronic amplification received in the infancy of sound motion pictures. Sound quality is approximately equal to that which electronic amplification had before the intense research by Bell Laboratories, Western Electric, and RCA. The immediate applications are Orthodox synagogue worship on the Sabbath, especially for the High Holiday overflow seating coverage, as well as locations in which the use of electricity may produce a possible hazard.

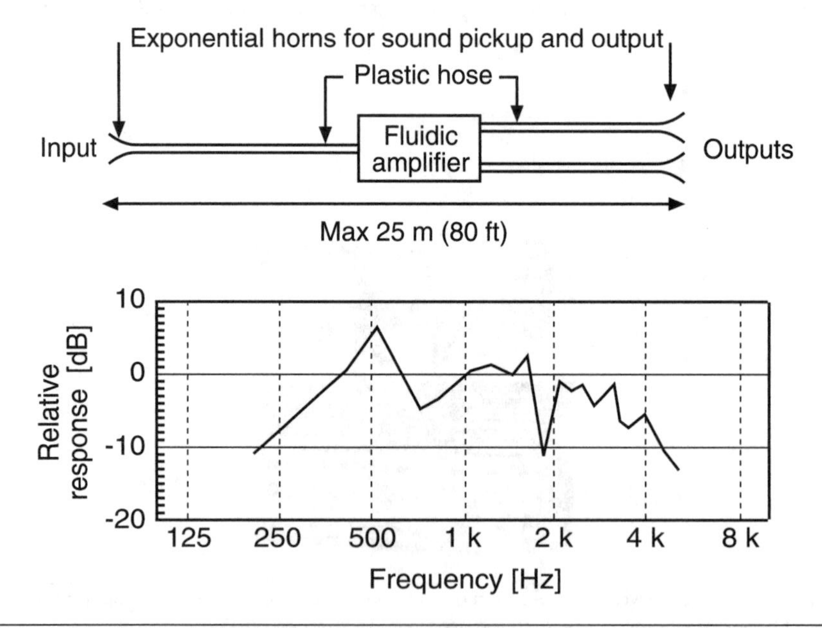

Figure 10.26 Air-only amplification. Experimental data from setup using 60 ft plastic hose and commercial horns (see Reference 10.11).

Note, however, that electronic amplification-employing devices in which current is varied without zero crossing, including condenser microphones that feed preamplifiers directly, has also received some Orthodox rabbinic approval. The principle of such a system is shown in Figure 10.27.

The availability of digital-signal processing, steered-line array loudspeakers can remove the problem of zero crossing in lines to the loudspeakers, and the availability of electrostatic loudspeakers, analogous to the condenser microphones, can eliminate the zero-crossing problem in the loudspeakers. Also, nanotube loudspeakers may be available that also require a current and voltage bias that removes the zero crossings. These developments should reduce further objections to Sabbath sound systems in synagogues.

Distributed Digital-signal Processing

The development of computer-steered array loudspeakers, with each small loudspeaker having its own attached power amplifier and signal processor, suggest applying the hardware to distributed systems for advances in quality and flexibility, including improved speech intelligibility. Each and every loudspeaker in a ceiling, chandelier, or pew-back system can have the optimum level and delay for its location rather than being somewhat compromised as part of a group of loudspeakers. Figure 10.28 compares a conventional distributed system and one with distributed digital signal processing.

Note the typical Latter Day Saints (Mormon) Stake and Ward Chapel and Cultural Hall combination shown in Figure 10.11. More than 1000 of these buildings have been built without any signal delay, with a signal delay for the cultural hall when used with the chapel, and with two or three delay zones, one for the cultural hall and one or two for the chapel with front loudspeakers without delay.

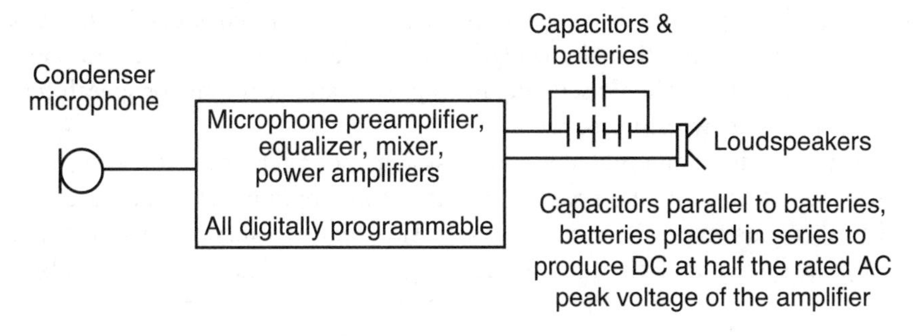

Figure 10.27 Single-sided amplification. A schematic of the basic principle of an electronic Shabbat-approved sound reinforcement system meeting Zomet Institute (Israel) requirements (see Reference 10.11).

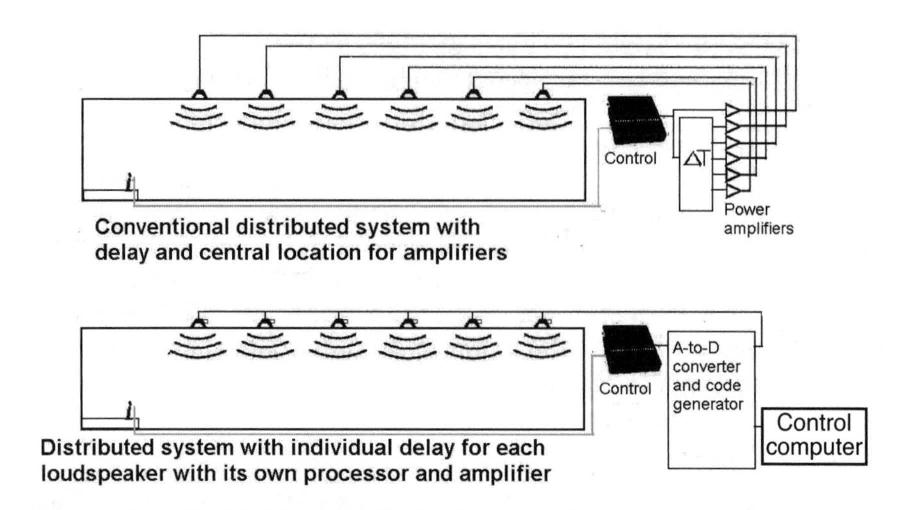

Figure 10.28 Regular delayed distributed system (top). Distributed digital-signal processing (bottom).

Depending on usage, this may be the most ideal application for distributed digital signal processing.

10.4 APPLICATIONS

10.4.1 Basic Reinforcement System

Some worship rooms rely entirely on natural voice projection, including many Orthodox synagogues and mosques; but those that only occasionally use voice reinforcement should do it in a way that is natural sounding and that contributes to intelligibility (qualities discussed at the beginning of this chapter). Most acoustical consultants agree that the central loudspeaker cluster approach to sound reinforcement is best when discussing intelligibility and economy for speech reinforcement in most worship spaces.

Cost considerations, not the sound system designer's best thinking, often govern the selection of the system. For large contemporary churches, a central loudspeaker cluster can be combined with side loudspeakers to form a flexible *broad-front* system that gives the sound designer for a particular event flexibility while using only the central cluster for greatest intelligibility for dialogue, the side loudspeakers for special stereophonic effects, amplification of massive choruses or on-stage orchestras, and the maximum level with minimum distortion for specific music uses.

A variation and supplementation on the broad-front system uses many small loudspeakers in the form of a horizontal line source in the orchestra pit railing for main floor coverage. In a relatively low-ceilinged space, a ceiling distributed system can provide intelligibility and naturalness if delay is properly applied. Delay is also a required application of pew-back or chair-back technology. Note that delay systems can improve directional realism for proscenium loudspeakers in certain cases also.

The Calvary Assembly of God, Orlando, Florida, pioneered the concept of having both a full broad-front central system and a complete distributed system in the same space, useful together or separately, to provide full redundancy in emergencies, and to envelop the congregant in high-quality amplified sound, much as a concert hall envelops a listener in natural sound. Pew-back and other types of systems are solutions for specific architectural situations; columns, lack of a location for central loudspeakers coupled with high ceilings and long reverberation time, for example.

The most traditional microphone pickup for speech reinforcement has been the use of microphones on stands or lectern bases. Today, the sound system designer can chose wireless systems which are essential for presenters who need to move while addressing the congregation. Other choices include boundary microphones and, for greater distances, presenter-to-microphone.

The large, evangelical, electronic churches frequently have all the systems described for theaters with great emphasis for on-stage, foldback coverage. The music director, the minister, and various performing musicians may request different mixes of the live, reinforced program. The emphasis placed on the quality and dynamic range of electronically reinforced sound is as great as on any other aspect of the church's efforts. TV and radio coverage is also extremely important.

In more traditional churches with strong music programs, the trend is to acoustical design favoring traditional liturgical music, and sound systems must ensure high intelligibility in reverberant spaces. Either directional loudspeaker systems that concentrate amplified sound into the congregational seating area or distributed systems with loudspeakers close to the sound-absorbing congregation and low amplified levels can be the solution. The specific techniques were discussed at the beginning of the chapter (see Figure 10.1). Combination systems are frequently employed, usually employing digital delay for parts of the system, and, frequently, wireless hearing assistance systems are included.

Many synagogues have a social hall added to the synagogue for High Holy Days. Usual practice is a separate distributed ceiling system added to the synagogue system via digital delay.

Orthodox synagogues rarely use reinforcement. The fluidics air system is in limited use, mainly for High Holy Day overflow use. Its very nature builds in the delay synchronizing sound in the overflow space with sound from the synagogue.

The single-sided approach that does not interrupt current in electronic reinforcement is built into the architecture of some important synagogues; extremely high-quality but unnoticeable reinforcement. The example most familiar to the authors uses a single button condenser microphone and four vertical line-source column loudspeaker systems with levels and equalization set for complete lack of awareness of any amplification. Additional speech clarity when the rabbi addresses the congregation is present with amplification from the lectern only. The architecture is traditional and the column loudspeakers are built into the wall on each side of the Aron HaKodesh, the Ark containing the Torah scrolls; one column on each side slightly above head height level for the men on the main floor and one on each side directly above head height level for the women in the three-sided balcony.

Mosque sound systems are basically simple, providing amplification only for the imam's address to the congregation and, separately, long and involved outdoor prerecorded calls to prayer from loudspeakers located high in the minaret. Past practice frequently involved the use of inexpensive but efficient paging horn loudspeakers for this application, but the trend has been to concert quality sound with two-way, wide-range loudspeakers suitable for music reproduction. The architectural design of the minaret must accommodate these theater-type loudspeakers in basic design. However, zoning in certain cities, and the desire of specific Muslim congregations not to disturb neighbors in any way, sometimes avoids the issue completely and the outdoor calls are simply not provided.

10.4.2 Archival Recording System

This system ensures a record of the worship services and the requirements for their use will determine whether the system should be single channel, stereo, or multichannel. Most often, the recording equipment is fed directly from the sound-reinforcement control equipment. When tapes are to be used for high-quality playback purposes such as radio broadcast or commercial recording release, separate microphones, normally suspended over the forward part of the audience, are used separately or with feeds from signals also used for reinforcement. In the best of circumstances, not often achieved, the recording system will have its own isolated control booth and control console, recording equipment, monitor amplifiers, and loudspeakers.

10.4.3 Monitoring and Paging Systems

These systems permit building staff, at their normal office locations or classrooms, to hear the worship services and choral rehearsals with interruptions by a manager to address specific people if necessary. Cry rooms and toddlers' rooms similarly require coverage. Specific requirements should, of course, add to or modify these suggestions. These systems are not operated generally other than a pager pushing a button. All level controls should be preset and the monitoring system should be left alone with most equipment rack-mounted. Loudspeaker coverage oftentimes is provided by built-in small cone loudspeakers in walls and ceilings, and loudspeaker-line level controls may be provided for specific dressing rooms and other spaces. The microphone pickup for program monitoring is similar to archival recording and includes the paging microphone, a push-to-talk close-talking communications type microphone, but high enough in quality to sound natural.

10.4.4 Surround and Electronic Reverberation Systems

Rooms may have acoustical deficiencies such as: (1) unsuitable ratios between direct, early reflected, and reverberant sound, (2) unsatisfactory early reflection patterns, and (3) short reverberation times that can be improved by surround and electronic reverberation enhancement systems. These systems are designed to complement particular acoustical conditions to add reverberation, liveliness, and a sense of surround for orchestral, choral, and organ music. The sound-field synthesis and reverberation enhancement systems are the two most common ones.

The sound-field synthesis systems are designed to provide the necessary *envelopmental* sound-field components such as early reflections from side walls and ceiling, as well as reverberation by using digital filters that can be designed to imitate a specific, desired room response. These systems are designed so that much of the feedback between loudspeakers and microphones is avoided (see Figure 10.29).

A sound-field synthesis system can generally be configured as shown in Figure 10.30. The signals from the microphones are sent to some form of digital-signal processing-based matrix system feeding several loudspeaker channels. It uses microphones and loudspeakers that are interconnected to form a complete matrix of channels in which signal processing may vary between channels. A matrixed system can be designed as an array loudspeaker system or as multiple

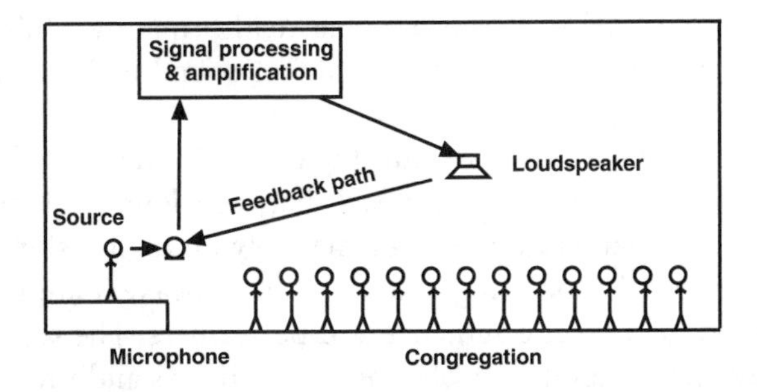

Figure 10.29 The feedback path between loudspeaker and microphone is practically unavoidable. Various methods are available to limit this problem.

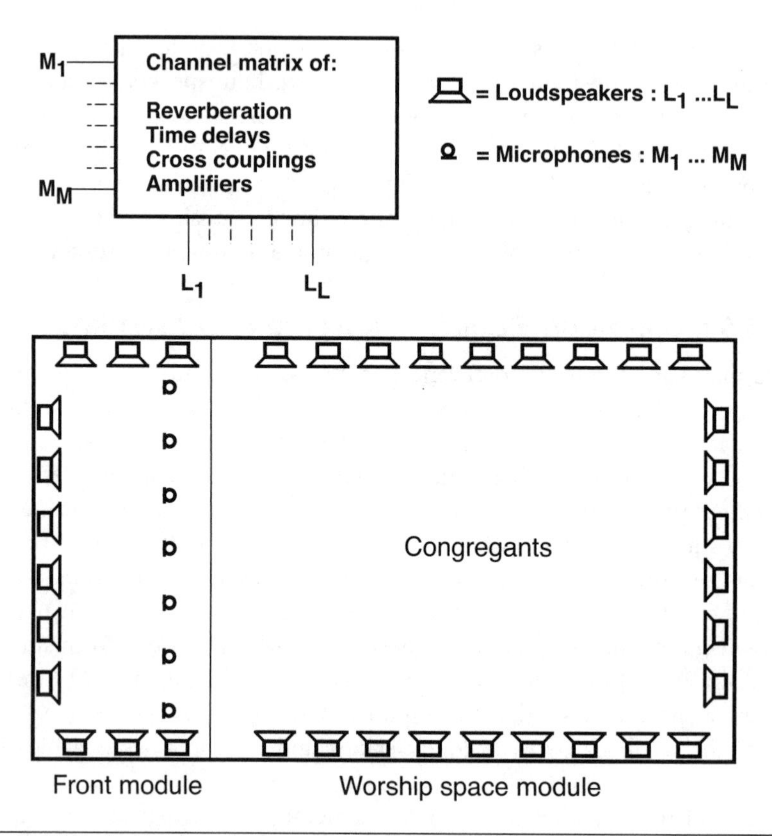

Figure 10.30 Typical loudspeaker arrangement for a reverberation enhancement system such as LARES, ACS, and others.

individual systems with many independent outputs. Digital-signal processing is used in this system.

Reverberation enhancement systems are designed to increase the reverberation time and sound level of the reverberant field primarily while, at the same time, having a negligible influence on the direct sound and the early-reflected sound.

Reverberation enhancement systems typically will use several microphones to pick up sound. They may use feedback between loudspeakers and microphones to generate reverberation or they may rely on digital-signal processing. Several forms of signal processing and many loudspeakers are used to obtain the desired sound field for the audience and performers. The signal-processing equipment may include level controls, equalizers, delays and reverberators, systems for the prevention of howl and ringing, as well as frequency-dividing and crossover networks.

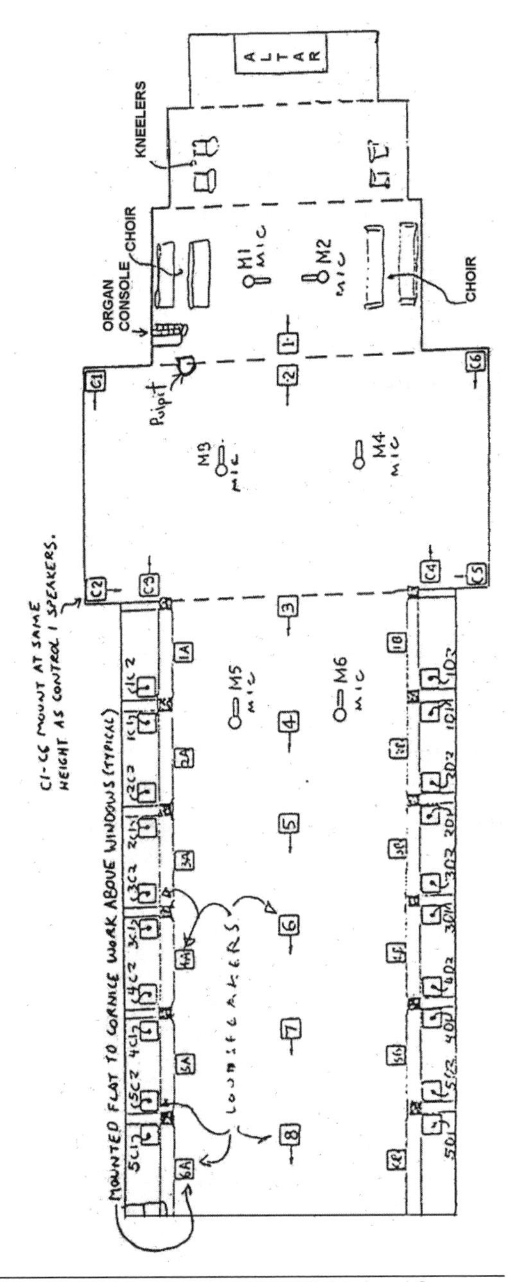

Figure 10.31 The LARES system is used at the Episcopal Church of St. Michael and St. George, St. Louis, MO (left) (photo: David L. Klepper). The plan with loudspeaker and microphone locations is from David Griesinger, inventor of the LARES system, and was used by the installing contractor (right) (see Reference 10.28).

Most modern commercial systems are combinations of sound-field synthesis and reverberation enhancement systems and use some form of subtle modulation to minimize feedback and ringing.

A system common in the United States is the LARES system (LARES Associates, Belmont, Massachusetts). Figure 10.31 shows the Episcopal Church of St. Michael and St. George, St. Louis, Missouri, where such a system is installed. For a general discussion of LARES, see Reference 10.2. The Church of Michael and St. George has a supplementary speech processing system as well (see Reference 10.19).

An example of the *wave-field synthesis* approach to natural sounding sound-field enhancement is the ACS system (Acoustic Control Systems, Garderen, The Netherlands) (see Figure 10.32). The most advanced of the ACS systems can be regarded as *holographic* systems since they use an array of microphones for sound pickup and an extensive, dense array of loudspeakers, surrounding the listening area (stage and audience areas may be covered). It is possible to simulate the reflected waves of a specific space much larger than the one in which the system is installed. For the system to work as intended, the transducers must be close (much less than a wavelength) and individually steered from the matrix.

The CARMEN reflector system (CSTB, Grenoble, France) uses integrated, small (wall mounted) microphone-loudspeaker units known as cells to pick up sound and retransmit it with suitable delay and reverberation characteristics (see Figure 10.33). Adaptive systems are used to control and eliminate the internal feedback in this beacon system. If a number of cells are used in a moderately sound-absorbing or sound-diffusing surface, one can create an illusion of a more remotely positioned sound-reflecting surface or object.

Figure 10.32 The ACS system is used at First Congregational Church of Battle Creek, MI. Consultants were Kirkegaard & Associates, Chicago, IL. Note the circular array of loudspeakers along the ceiling (photo: ACS).

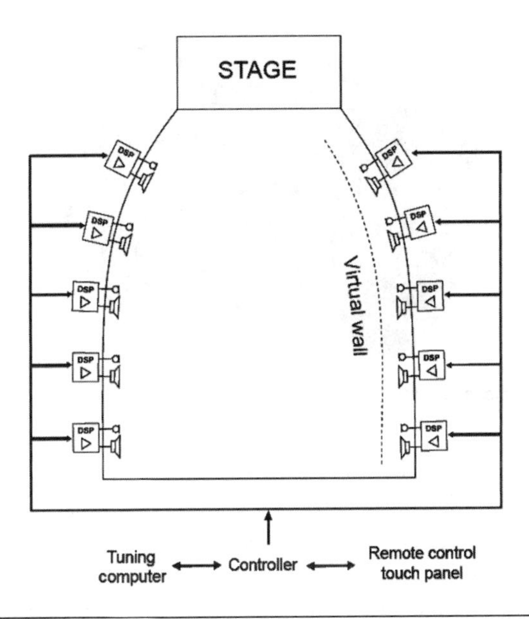

Figure 10.33 The CARMEN system uses individual, reverberant cells that respond to the incoming sound by sending out a reverberated signal, creating virtual walls (adapted from information provided by CSTB, Grenoble, France).

10.4.5 Acoustical Envelope or Stage Communication System

This system, found in the large mega-churches featuring contemporary music, allows singers on the stage to hear a pit orchestra better, all performers on a stage to hear themselves and each other better, and possibly allows clergy and others on a stage to hear more effectively. This system may share components and loudspeaker lines with a theatrical sound-effects playback system—a subject beyond the scope of this book. Special *slant-faced* portable loudspeaker systems have been developed for these applications (see Figure 10.34). Popular music stars frequently require amplified sound to be fed back to them and a separate control console and control operator are assigned for this purpose.

10.4.6 Hearing Assistance and Simultaneous Translation

Many facilities now are required by law to provide assistance to the hearing impaired, and this can be done by wired systems with earphone jacks at the seats, by FM wireless or infrared transmission and portable receivers with earphones and batteries that are recharged between performances. If multichannel, such systems can also serve as simultaneous translation systems. One or more translators can listen to the live performance via earphones in translation booths, viewing the performance through windows and then speak into a local microphone.

Figure 10.34 A JBL 7RX112M slant-faced stage monitor loudspeaker (photo: courtesy of JBL/ Harman International).

In a small church, synagogue, or mosque, the hearing assistance system may, in fact, be the only reinforcement system requiring microphone pickup of the live performance. But simultaneous translation may also be a requirement for multi-language congregations.

10.4.7 Production Communications

The huge mega-churches may require communication during rehearsals and services between a stage manager and all the technical and special effects people or the lighting and sound control operators. These may be single-, two-, or multi-channel systems. Certain stations may be built-in, but many will be worn by the people such as the wired belt-pack systems that are typically used today. Batteries may be recharged between shows. Some churches use completely wireless systems.

10.5 SYSTEM PLANNING AND IMPLEMENTATION

10.5.1 Design

The design of a sound system consists of several steps, beginning with the basic decisions of what the system is to be used for, and then extending the plans to the acoustical, architectural, and electronic design.

Programming

The first issues to be addressed in the design of a sound system involve answering the questions:

a. Who (or what) is to communicate to whom?

b. Where is the sound source and where are the listeners?
c. Is the system to provide intelligibility? reverberation and ambience?
d. How important are directional realism and a natural sound quality?

These decisions can be incorporated into a set of functional requirements that explain what the sound system is supposed to do, including requirements such as recording, playback, and perform foldback (or on-platform monitoring), as well as the nature of the audience coverage.

Performance requirements are then determined with regard to sound levels needed, frequency response, clarity as measured by articulation-loss-of-consonants or early-to-late reverberant energy ratio (see Reference 10.29). These might require some modification later in the design process as a result of budget requirements and/or architectural limitations. Figure 10.35 shows a flow diagram for system planning of a typical traditional church.

Acoustical/Architectural Design

The designer must then determine the architectural limitations on the design of the system and the acoustical characteristics of the spaces served. Decisions then will be made on the type of loudspeaker system or systems that can provide the coverage in the best way to meet these requirements. At this point, the detailed design of the loudspeaker system can begin. That involves detailed calculations and, today, the use of computer programs in all but the simplest design problems. Microphone requirements should be studied, microphone receptacles located, and the complete system analyzed from the perspective of gain-before-feedback, coverage uniformity, and power amplifier output requirements. On occasion, some redesign will be necessary based on the need to meet architectural and/or budget limitations (see Reference 10.30).

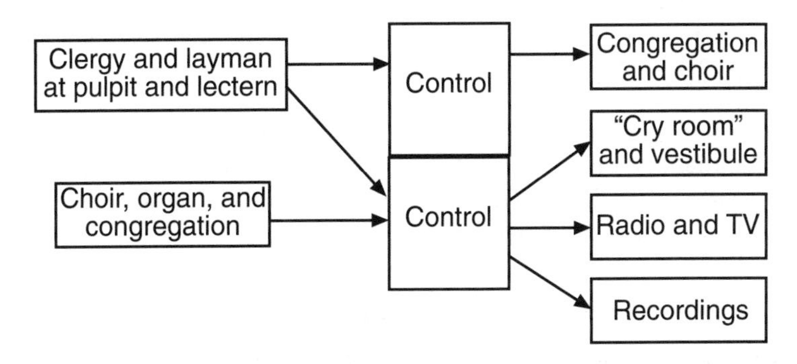

Figure 10.35 Flow diagram for system planning of a typical traditional church. Conservative or Classic Reform synagogues may be similar.

Electronic Design and Cost Estimating

The electronic design of the system must be based on the signal flow (routing) requirements and the needed amplifier power at each of the loudspeaker terminals. This leads to the selection of preamplification and control equipment, equalizers, other signal-processing equipment, and power amplifiers. The heat load of the system must be determined to provide the mechanical engineer with the air conditioning requirements for the rooms in which electronic equipment is located. The detailed design then proceeds to rack layouts, conduit and wiring diagrams, control arrangements (often involving considerable *human engineering*), and a final selection of the equipment for the entire system. A draft functional diagram should be prepared showing all electronic, electro-acoustics, and recording/playback equipment and connections. Two examples of a functional diagram are shown in Figures 10.36 and 10.37. The next step is a cost estimate that may lead to some redesign and some rethinking of the priorities in the functional requirements.

Specifications

If a design-build contractor negotiates directly with the owner, the specification step may be omitted, but if competitive bidding is required then detailed specifications will be necessary (see Reference 10.30). In some cases, these may be in outline form and the bidders restricted to those that are known to have a proven track record. Written specifications usually include:

1. General legal requirements such as insurance
2. Collaboration with other trades
3. Functional and performance requirements
4. Outline design of the system
5. Equipment listed by manufacturers and type number or by technical requirements or both
6. Installation techniques
7. Tests and adjustments to be performed on the completed system
8. Service and guarantee requirements

Drawings usually include a functional or line diagram showing the interconnection of equipment (see Figure 10.37), the location and orientation of all loudspeakers, and possibly details of hardware such as receptacle boxes, custom control panels, and any custom hardware.

Bidding, Contractor Selection, and Supervision

When several contractors bid for one sound system project, a consultant often has to choose one on the basis of price, conformance with specifications, and

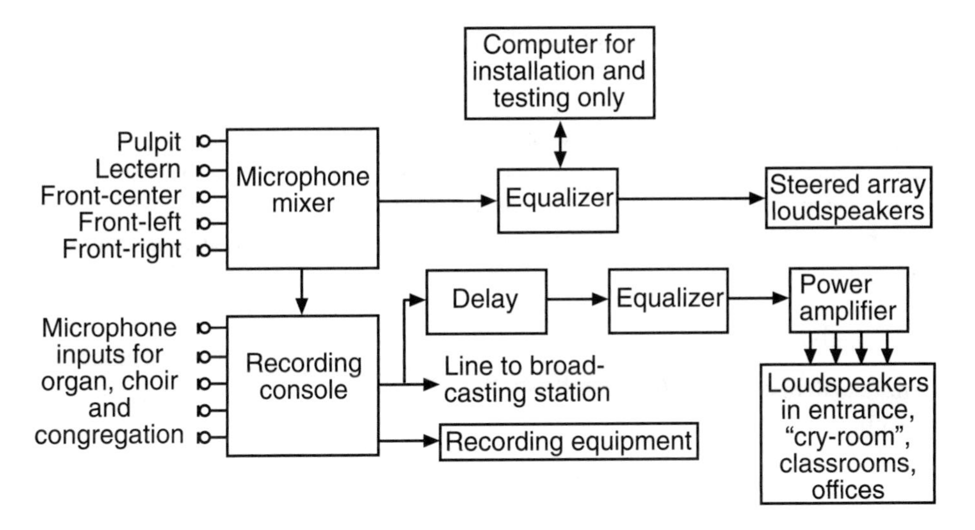

Figure 10.36 Functional diagram based on Figure 10.35. Signal delay for auxiliary loudspeakers is required to prevent entrance area loudspeakers as a pre-echo source to sound from the worship space.

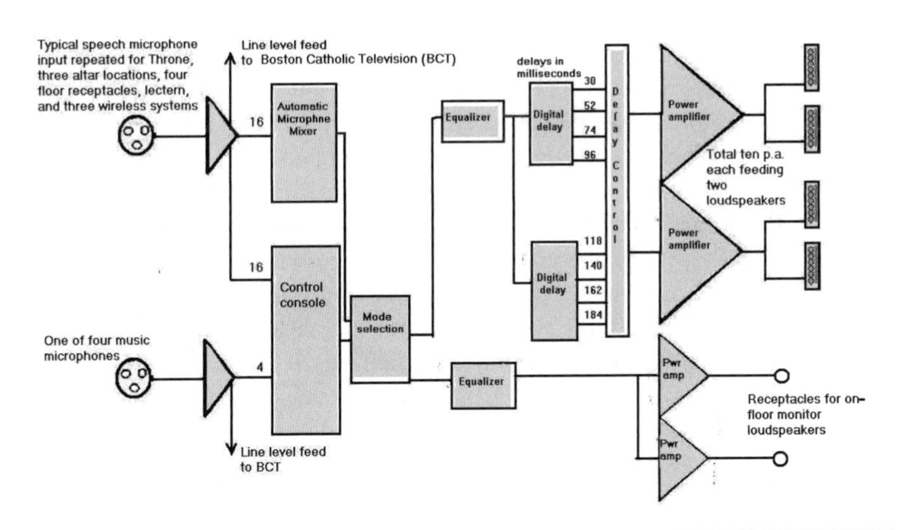

Figure 10.37 Functional diagrams based on Figure 10.35 for Holy Cross Cathedral system of Figure 10.10.

their track record for completing similar projects. This may be the time to discuss alternatives suggested by the chosen contractor as to the specific equipment specified by the designer, or this process may continue through the supervision phase. Supervision will involve the review of shop drawings that cover installation details, site inspections when required, and generally being responsive to any questions that arise on the project. The contractor may prepare detailed wiring diagrams for the installation. Generally, one qualified engineer should supervise an installation from start to finish.

Installation Practices

The usual installation for high-quality sound systems includes (see Reference 10.30):

1. Fastenings and supports for all fixed equipment and components that provide a safety factor of three or better.
2. Precautions to guard against electromagnetic and electrostatic hum pick-up, particularly where the use of dimming equipment in the lighting system is normal.
3. Precautions to ensure adequate ventilation of the equipment and, of course, to ensure the safety of users and operators of the system, including any possible hazard.
4. Separate conduits for microphone-level circuits (below −20 dBm), line-level circuits (−20 to +30 dBm), and loudspeaker circuits (+30 dBm and above), switching circuits, and power circuits, when possible. Where absolutely necessary, the first two and last three circuits can be run together, but shielding must be carefully thought-out for the individual cables. The individual signal lines should, in any case, be fully insulated from each other and from their conduit, and the conduits mechanically and electrically connected to receptacle boxes and to the building's electrical ground connection.
5. Splices should be avoided for all signal lines in conduits; microphone line shields grounded at the patch panel frame or input jacks, whichever is applicable; and other shields grounded at the inputs of control and processing equipment, or power amplifier inputs, as applicable.
6. All audio grounds in equipment racks should be connected to a common point and the racks grounded to the grounding point.
7. Wiring joints and connections should be made with rosin-core solder or with approved mechanical connections.
8. Equipment should be installed neatly, true to line and level, plumb, and square. All controls, receptacles, and cable terminations should be clearly, logically, and neatly marked. The contractor should coop-

erate with other trades for good installation progress, acceptable final appearance, and clean job sites free from damage (includes wiring), marks, and blemishes.

Shop Drawings

The contractor may use the contract drawings as a basis for shop drawings or may prepare originals. In any case, the shop drawings should go beyond the contract drawings to show the details of all cables and connections, including color coding and shields, screw and other terminals numbers and labels, rack layout, normal control settings, descriptions of conduit runs and terminations, details of loudspeaker mounting arrangements and microphone receptacle boxes, a list of all equipment down to screws and connectors not part of the major equipment that makes up the major portion of the list, and all equipment input and output impedances.

Adjustment, Equalization, and Final Approval

The sound system contractor should perform the following tests on a completed sound system:

1. Impedance of each loudspeaker line.
2. Electronic hum and noise electrical signal level of each output channel with normal gain settings and each input terminated with a resistor matching the normal microphone input impedance.
3. Total harmonic distortion at two frequencies, and possibly the intermodulation distortion of each output channel with the signal routed through the entire appropriate signal path and the use of normal gain settings.
4. Tests on the complete system, including the use of input equipment such as microphones to ensure complete freedom from parasitic oscillation, radio-frequency pick-up, hum, and noise. Dimmer lighting-control systems should probably be operated through their entire range as part of these tests.
5. Tests to ensure in-polarity operation of all loudspeakers that should operate, including all portions of distributed and line-source systems, adjacent loudspeakers in multi-loudspeaker arrays, as well as portable loudspeakers used for sound effects and foldback. There may be specific applications in which a null might be created at a microphone location where out-of-polarity operation may be preferred.
6. Tests on the complete system, particularly the loudspeaker systems, to ensure freedom from buzzes, rattles, and distortion. Often the best

test is the playback of recorded music at the highest reasonable levels over the system.

Coverage uniformity and frequency response are most frequently measured with a pink-noise signal generator and real-time analyzer (usually one-third-octave-band for frequency response and octave-band for coverage), but modern measurement techniques such as time-energy-frequency measurements can often provide additional information, including separation of the direct sound energy from the reverberant field. Also, the technique *source independent measurement*, developed by Meyer Sound Laboratories, can provide frequency response data during an actual speech or music program and, thus, has a particular value with the adjustment of portable systems or systems used in spaces in which the acoustical balance across the frequency range varies widely with changes in occupancy. All these measurements can be used during the adjustment of balances between various loudspeakers, adjustment of crossover networks, and the setting of all equalizer controls (see Reference 10.30).

Delay settings, when required, usually are arrived at by use of the *time-energy-frequency* (*TEF*) measurements referred to earlier, subjective responses to clicks played over the system, or program material, or some combination of two or three of these techniques. The LGMarshall/Techron Program, *RASTI* measurements, and other direct reading of intelligibility can speedup the delay setting process and increase accuracy (see Reference 10.31 and 10.32).

Finally, a written test report with all the data can provide a written record of a system's success and act as a guide to future maintenance, adjustment, and possibly expansion or alteration to meet new needs. The contractor and designer should work closely with the sound system operator during the first events, and the operator and/or owner should be ready to contact the contractor and/or designer in the event of any question or problem. Checkups should be performed regularly, completing the earlier steps toward an excellent system with a long life expectancy (see Reference 10.33).

PART II

SYNAGOGUES

S.1 HISTORY

Of the three major monotheistic faiths, Judaism is the oldest, thus, synagogue architecture predates mosque and church architecture. Ancient Jewish worship is well defined in the *Torah* (the first five books of both the Jewish and Christian Bibles) and involved the desert Tabernacle and Jerusalem Temple, both defined in the Torah and other Jewish sacred texts. The Tabernacle had a length that was three times its width.

The basic design of Solomon's Temple, the first temple built in Jerusalem, is well known. This temple was destroyed in 586 BCE when the Jewish kingdom was overtaken by the Babylonians and the Jews were exiled into Babylonian captivity. The psalms of that time mention the use of musical instruments in the temple; the organ, the lyre, and, of course, the priests' songs. The second temple was built on the Jews' return from exile around 516 BCE and existed for a considerable time. A model of the second temple can be seen in Figure S.1.

The Romans destroyed Jerusalem and the second temple ending the great Jewish revolt that began in 66 CE. The Babylonian exile had made it necessary to devise new forms of worship in which prayer replaced sacrifices. After the destruction of the temple, the rabbis forbade the use of musical instruments in Jewish worship on the Sabbath or the High Holy days. The shofar is the only instrument that is allowed in Orthodox synagogues, except for the voice. It is used throughout the month of Elul preceding the new year (Rosh Hashanah) and through Yom Kippur (see Reference S.1).

The synagogue was introduced as a meeting place, study hall, and worship place. The term synagogue comes from a Greek word meaning assembly. In

Figure S.1 The model of the Second Jerusalem Temple at the Israel Museum, Jerusalem, Israel (photo: Mendel Kleiner).

Hebrew, the term is Bet Knesset. which also means assembly house. Few remains have been found of the early synagogues, but some of the many synagogues built in the first few centuries CE have been found. Excavations of such early synagogues generally indicate simple, rectangular halls of worship based on the Roman basilica shape (see Reference S.2).

Figure S.2a shows the synagogue site at Gamla in the Golan Heights, Israel, which was excavated 1976-1989 CE. Gamla is the best preserved example of a Jewish city that flourished in the last two centuries of the Second Temple Period and is one of the few examples of a battle site of the first century CE in the Roman Empire that was left as it was abandoned. The siege and battle of Gamla are described in detail by Flavius Josephus (*History of the Jewish War: IV:I; §1-83*). Figure S.2b shows a suggested reconstruction of the Gamla synagogue by the Israel Antiquities Authority.

The synagogues in the Galilee and Golan Heights area are assumed to have had flat ceilings generally because wood was scarce in this region. Such synagogues typically have three rows of columns forming a main floor, nave, and two side galleries. A U-shaped balcony was held by the pillars. The building materials—stone, wood, and mosaic—were all hard substances, and sound absorption was provided by light openings, the ceiling, and the congregants. Typically the discovered synagogues that are found are small and have a floor area of less than 200 m^2 (around 2200 ft^2).

An example of a synagogue that was common five centuries later is the one at Kfar Nahum on the north shore of the Sea of Galilee in Israel (see Figure S.2c). This synagogue is 20 by 30 m and somewhat larger and was constructed

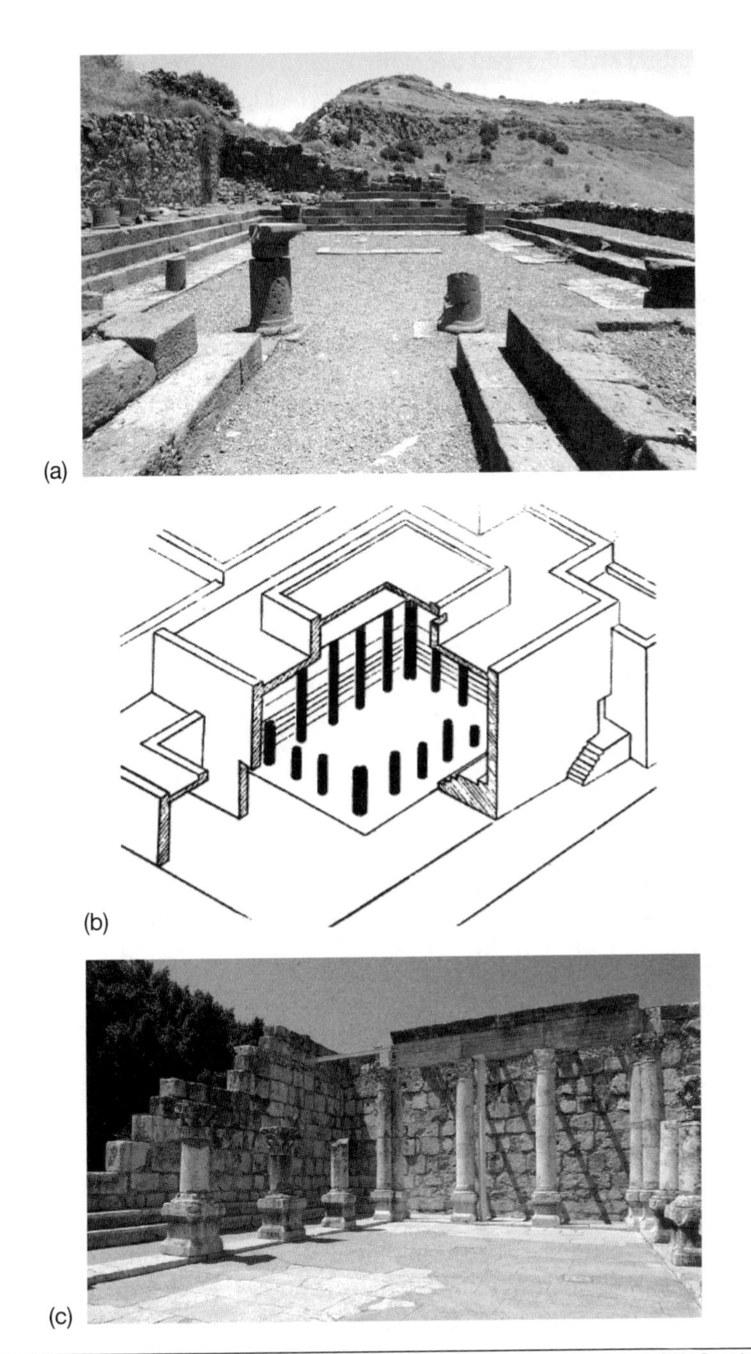

Figure S.2 (a) The Gamla Synagogue was destroyed by the Romans in 68 CE, Gamla, Golan Heights, Israel (photo: Mendel Kleiner); (b) Suggested reconstruction of the Gamla synagogue by the Israel Antiquities Authority; (c) Kfar Nahum Synagogue, Kfar Nahum, Israel (photo: Mendel Kleiner).

in an architectural style that had a lengthy tradition in Roman Syria. Frequently, there are two or three entrances to these synagogues but no antechambers. Entering through the main door, one would see at the far end in the direction of Jerusalem, a richly embroidered curtain, the parochet. Behind the curtain is the Aron Ha-Kodesh, the recess and cabinet in which the Holy Scriptures—the Torah Scrolls—are kept (see Reference S.3).

When the second temple was razed by the Romans, the synagogues remained as places for worship and the prayer service replaced the offerings. Because of the Roman expulsion, there were then more Jews outside of the Land of Israel than actually in Israel. In the small synagogues in the Land of Israel, the speech intelligibility should have been quite high because of the hard, sound-reflecting materials used in the construction. However, some of the synagogues were large. The Talmud, the great book of Jewish commentary written while the Jews were in Babylonian exile, tells of the Great Synagogue in Alexandria. Due to the substantial size, speech intelligibility was so low that a flag had to be waved to signal the most distant congregants that a prayer response, such as Amen, was due (see Reference S.1).

In the first centuries CE, the synagogues also took on new functions as hospices and providing chambers for needy travelers. Some synagogues were converted to churches as the Jews were expelled from various countries. Such a synagogue is shown on the far left in Figure S.3. This Toledo, Spain, synagogue became a church and is now a museum. It uses the key symbol that is prevalent in both mosques and near-Eastern synagogues, including modern day ones such as the Jerusalem Le-Benyamin Amar Congregation also shown in Figure S.3.

The synagogues that were built in the Middle Ages were limited in size because the Church devised regulations against them being higher than the surrounding Christian buildings. Additionally, the Ghetto system, in which the Jews were forced to live, limited the space available. Examples of such synagogues can be found in Italy, for example, in Ferrara. Similar regulations also existed in Muslim lands regarding synagogue height. A specific Jewish architectural style was never developed because Jews were always a minority in the countries where they dwelled and they used local builders and architects. The rectangular style remained the common one.

As long as the rooms were narrow and the ceiling height limited, the speech intelligibility was adequate. This was especially true because musical instruments were not allowed in the service. In addition, literacy was high and most congregants were able to lead a service and, therefore, did not need many cues. From the viewpoint of speech intelligibility, the most religiously important part was to be able to hear every word of the reading of the Torah. This was ensured by placing the bimah, the platform from which the Torah is read, at the center of the nave floor raised as much as eight steps above the floor.

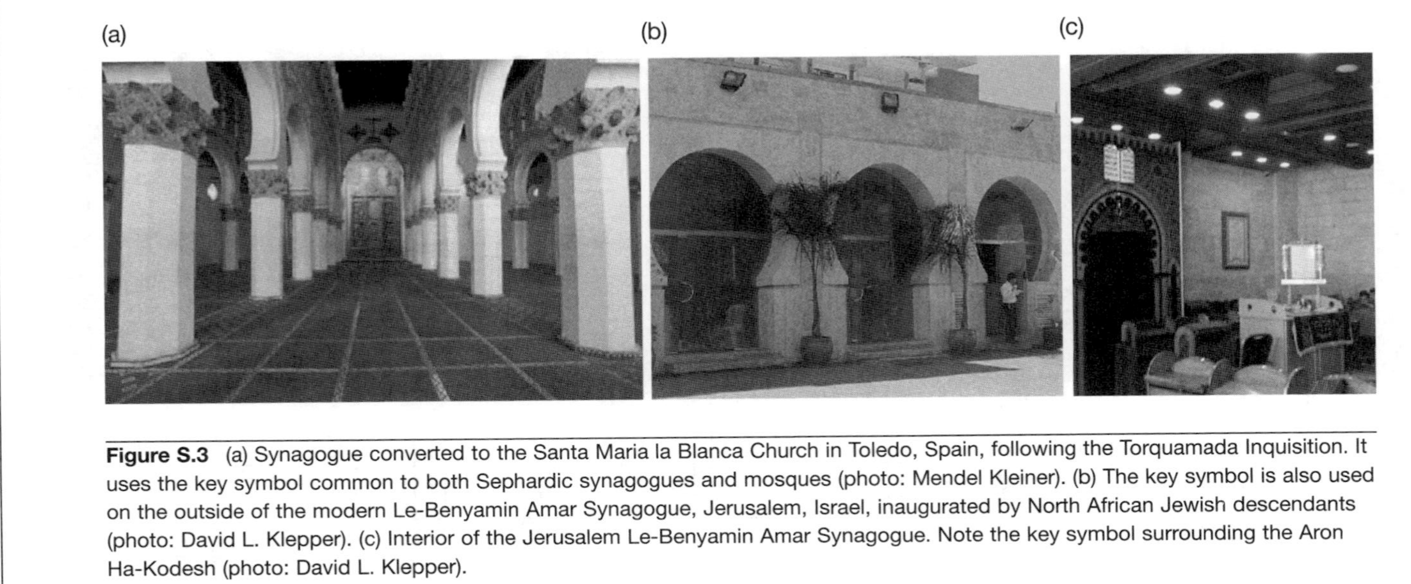

Figure S.3 (a) Synagogue converted to the Santa Maria la Blanca Church in Toledo, Spain, following the Torquamada Inquisition. It uses the key symbol common to both Sephardic synagogues and mosques (photo: Mendel Kleiner). (b) The key symbol is also used on the outside of the modern Le-Benyamin Amar Synagogue, Jerusalem, Israel, inaugurated by North African Jewish descendants (photo: David L. Klepper). (c) Interior of the Jerusalem Le-Benyamin Amar Synagogue. Note the key symbol surrounding the Aron Ha-Kodesh (photo: David L. Klepper).

S.2 MODERN SYNAGOGUE ARCHITECTURE

The traditional or Orthodox prayer service is based on the laws of the Torah and on biblical exegesis that have developed over the last two millennia. The written word carries the message. Most current worshippers following the Orthodox tradition will be reasonably familiar with the prayer book, the Hebrew text, and

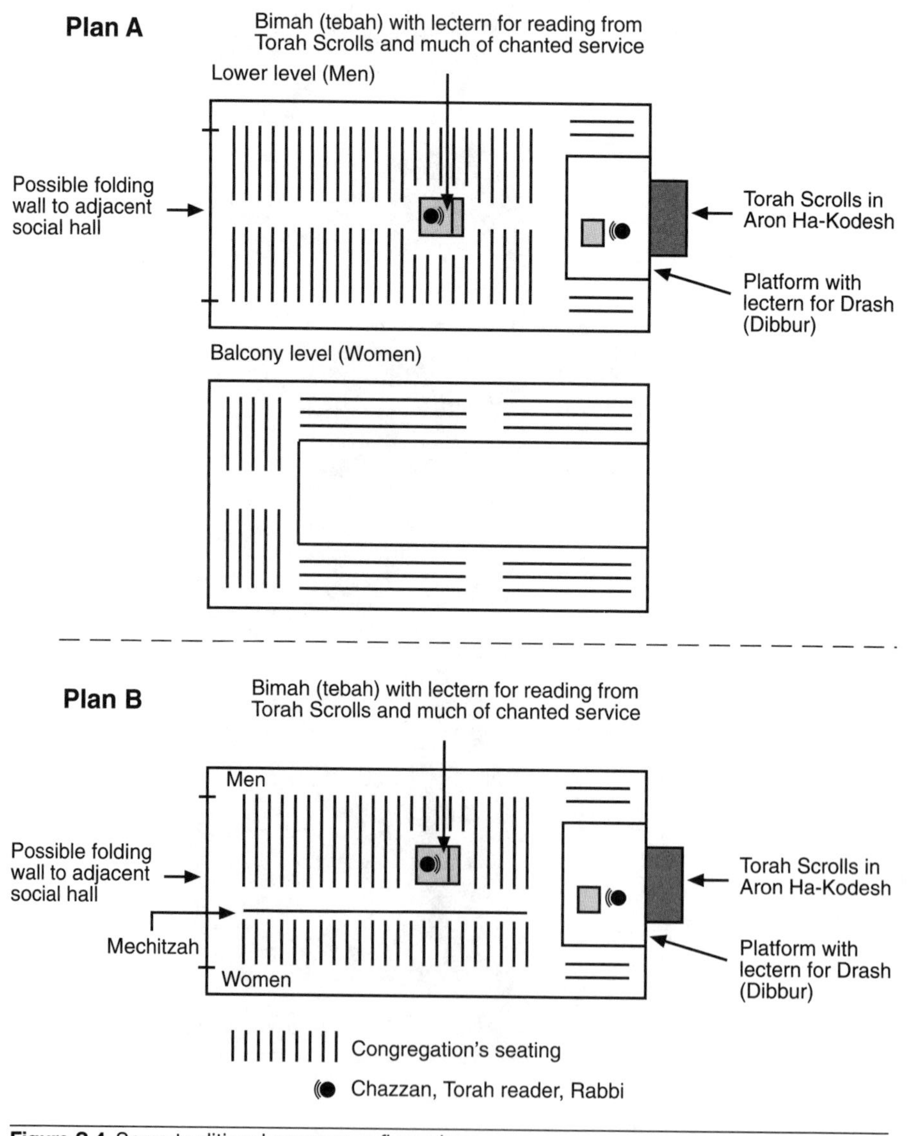

Figure S.4 Some traditional synagogue floor plans.

its translation. The service consists of communal participation in the reading of the prayer book text and in the reading of sections of the Torah, particularly on the Sabbath. There will be a prayer leader or precentor known as a chazzan, who is oftentimes a layman with the required knowledge and a good voice. At times, there is a small choir, but no musical instruments, neither mechanical nor electrical. These are not accepted on the Sabbath. This requires music in Orthodox synagogues to be vocal, song by the chazzan, a choir, and the congregants (see Reference S.4).

The basic floor plan of many modern, Orthodox synagogues is directly inherited from the early designs mentioned previously and shown in Figure S.4. Corresponding synagogue sections are indicated in Figure S.5. The chazzan leads the service by both continuous and interrupted praying. In most portions of the service, the chazzan will only read the last few sentences of each prayer outloud and acts a synchronizer for the congregation of worshippers. The interrupted prayer does not require excellent speech intelligibility so there is rarely a problem in having the chazzan lead the prayer from a position in front of the congregation, close to the Ark, and facing away from the congregants (see References S.1, S.4).

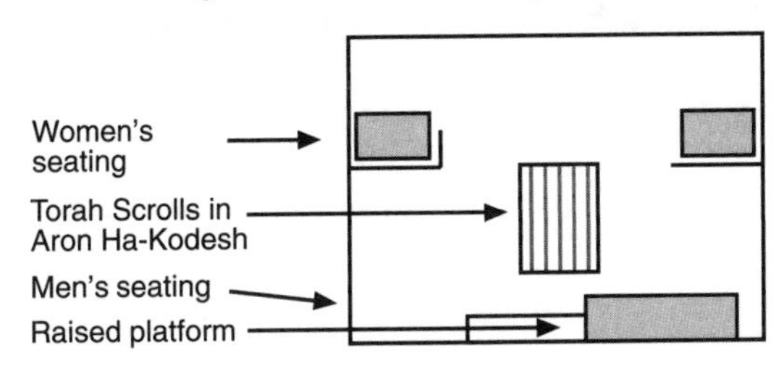

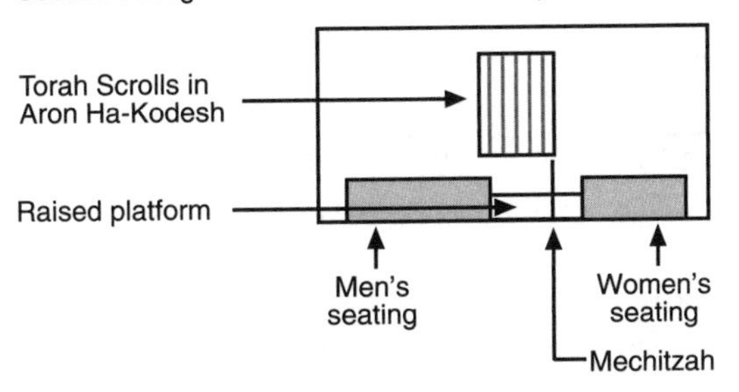

Figure S.5 Some traditional synagogue sections (see also Figure S.4).

On the occasions when the Torah is read, including Mondays, Thursdays, the Sabbath, and on many other occasions, the Torah Scrolls are carried by a procession from the raised platform on which the Aron Ha-Kodesh—Ark—is kept to the bimah, the raised platform at the center of the nave. The Aron Ha-Kodesh is nearly always on the eastern wall. The Torah passages are read by someone who must be well versed in the art of the chant, which can be regarded as a form of song. Since the voice of the Torah reader is directed toward the parochet and Ark, and the

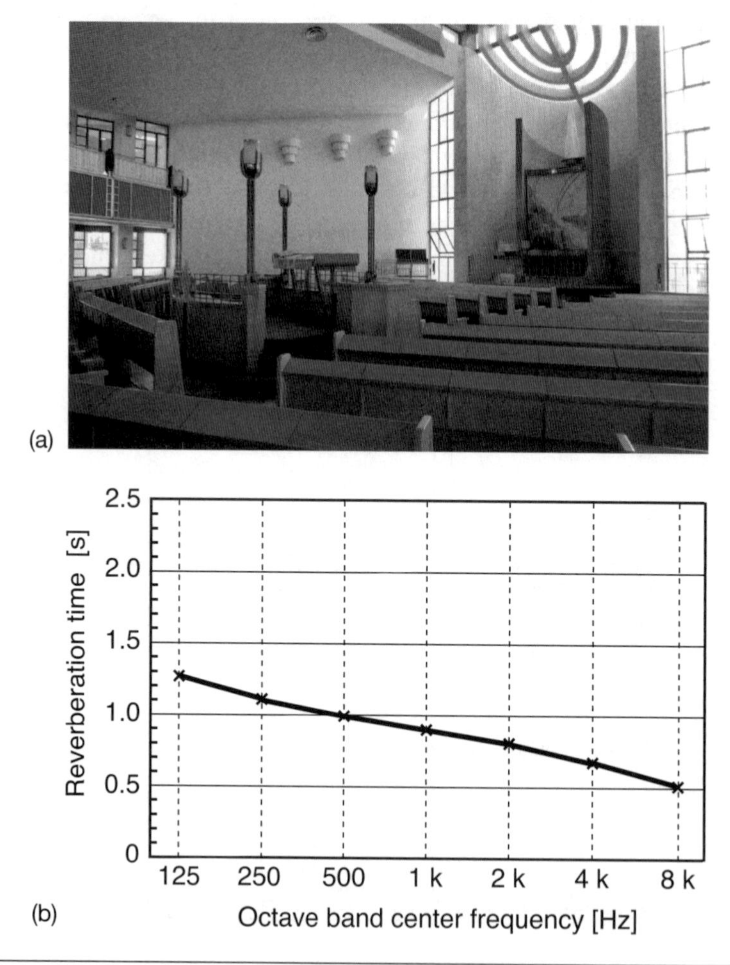

Figure S.6 (a) A modern Orthodox Ashkenazic synagogue in Alon Shvut, Israel, using a hexagonal plan with balconies (photo: Mendel Kleiner). (b) Reverberation time ($RT_{-5,-25}$) for the synagogue.

congregants are required to hear and perceive every word being read, there is a practical limit to how large a synagogue can be.

The rabbi will expand on some issue of the Torah while facing the congregation; this presentation is called the *Drash*. If the room is fairly small, the speech intelligibility will be high provided the reverberation time is sufficiently short.

Figure S.6a shows an example of a modern Orthodox Ashkenazic synagogue in Alon Shvut, Israel, according to Plan A in Figures S.4 and S.5 except that it uses a hexagonal plan. The balconies are low and have horizontal undersides which make it difficult for sound to reach. The reverberation time in this synagogue is short (see Figure S.6b) primarily due to the ceiling that has been treated with acoustic plaster. Additionally, seat upholstering and windows are factors.

The joint Sephardic synagogue next door, shown in Figure S.7, also has a hexagonal floor plan but no balconies which provides better intelligibility.

Another example of a modern Orthodox Ashkenazic synagogue according to Plan A in Figures S.4 and S.5, also using a hexagonal plan, is Jerusalem's Bet Knesset Gadol, the main Orthodox synagogue in Jerusalem shown in Figure S.7a. Because of its large size, the reverberation time is relatively long as shown in Figure S.7b.

Figure S.7 In most orthodox Jewish congregations, such as the Sephardic one in Alon Shvut, Israel, the chazzan's voice is directed away from the worshippers and toward the Aron Ha-Kodesh. Note the mechitzah on the right (photo: Mendel Kleiner).

S.3 SEPARATION OF SEXES: THE MECHITZAH

Orthodox worship requires that men and women pray separated physically. The upper section in Figure S.5 shows one form of the traditional orthodox layout in which women pray on the upper level just as they did in the synagogues of antiquity. Other plans, typical of small synagogues, will contain a women's section on the lower level next to the men but separated by an opaque screen or something similar such as illustrated in the lower section of Figure S.5. This layout is common in modern orthodox synagogues. Here, men and women pray separated by

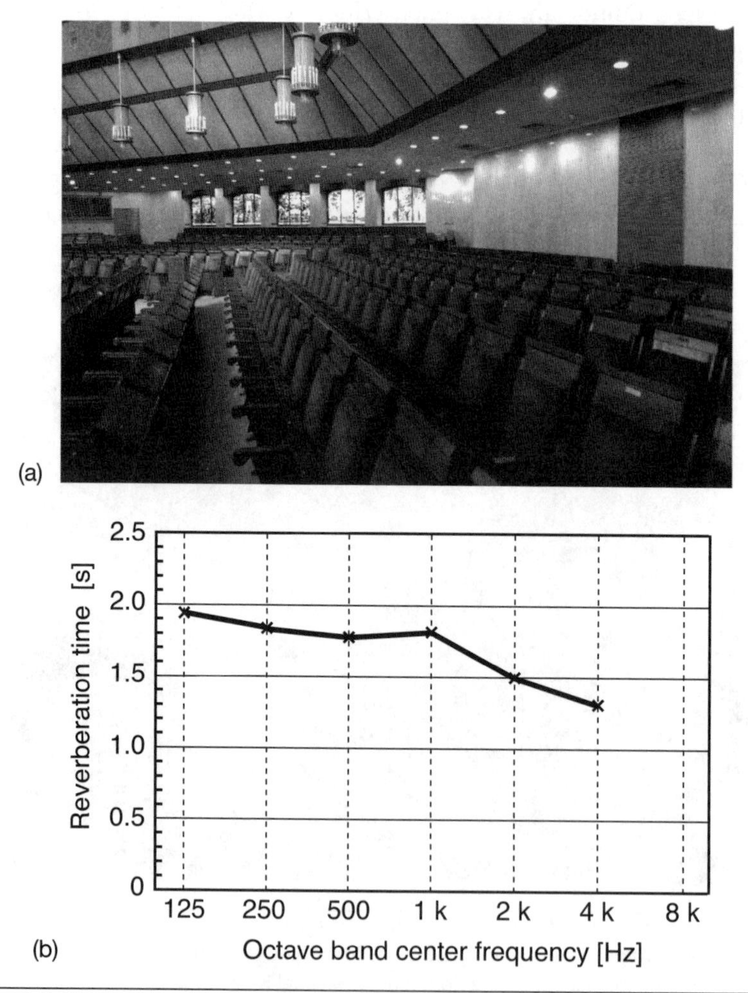

Figure S.8 (a) The Bet Knesset Gadol in Jerusalem, Israel, also uses a hexagonal floor plan (photo: Mendel Kleiner). (b) Reverberation time ($RT_{-5,-25}$) for the synagogue.

a screen which is called a mechitzah (pl. mechitzot). The right side of Figure S.7 shows one form of mechitzah. Note also the quite different mechitzot shown in the photos in Figures 7.27 and S.19.

Men will pray on the lower level, because they will need to participate in the reading of the Torah. Since reverberation time is fundamentally determined by room volume, the ceiling height is of great importance. Typically, most of the worshippers will be men. In principle, men are obliged to attend services both in the morning and in the evening. Frequently, the balconies will be empty and there will be minimal sound absorption as a result. The reverberation curves will be dual-slope due to the late reverberation coming from the volume above the balconies. Some synagogues do use pew cushions that control this phenomenon.

S.4 SMALL SYNAGOGUES

Figure S.9a shows an example of the small synagogue oftentimes called a *Shtiebel*, which means "small room" in Yiddish—the vernacular language of east-European

(a)

(b)

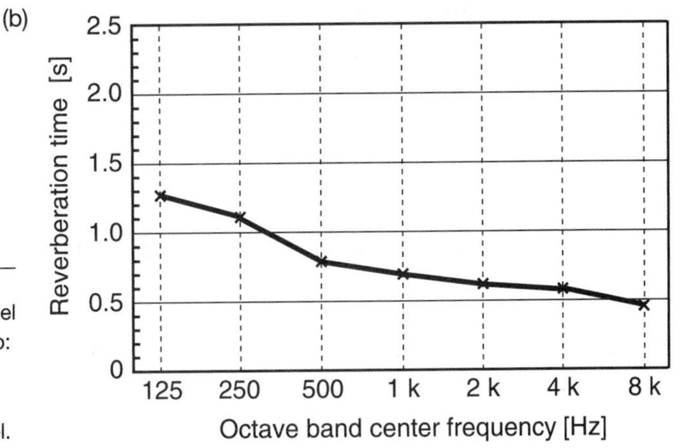

Figure S.9 (a) A small synagogue in the shtiebel style, Efrat, Israel (photo: Mendel Kleiner). (b) Reverberation time ($RT_{-5,-25}$) for the shtiebel.

Figure S.10 Touro Synagogue, Newport, RI (photo: congregation Jeshuat Israel).

Jewry before World War II. The shtiebel commonly has a floor area of only 20 to 30 m^2. These small synagogues may remind the occasional visitor of the Calvinist worship spaces. The room is usually not decorated and the seating may not be upholstered. Because of this, the main sources of sound absorption and scattering will be the worshippers, books, and book shelves. Scattering and sound absorption is generally high in relationship to volume because of the low ceiling height that leads to short reverberation times as shown in Figure S.9b. Unfortunately, the presence of sound absorbing ceiling tiles in some shtiebels today leads to difficulties in communication during crowded events.

The oldest synagogue building in the United States, the Touro Synagogue in Newport, Rhode Island, shown in Figure S.10, also fits these characteristics acoustically (without sound absorption), even though it is certainly also a monument (see Reference S.6).

S.5 SYNAGOGUES AS STUDY HALLS: BEIT HA-MIDRASH

There are also large prayer rooms such as the ones shown in Figure S.11 (see Reference S.7, S.8). One should note the absence of pews that are usually replaced by tables and chairs so that the room can be effectively used for study as well as prayer. Walls are bare except for pictures, posters, and bookshelves. Because of

Figure S.11 Two synagogue study halls. Congregation Young Israel of the Main Line, Bala Cynwyd, PA (left) (photo: congregation Young Israel). Large hall at the former Beit Midrash of Yeshivat Beit Orot, Mount of Olives, Jerusalem (right) (photo: by the Yeshiva).

the small volume, the reverberation time will be fairly short. Generally, speech intelligibility is high except when (a) the ceiling has been excessively covered by acoustic tile, (b) window-mounted or other noisy air-conditioners are used, or (c) movement of chairs results in screeches from the contact of metal legs on stone, concrete, or ceramic floors. This is a problem easily solved by rubber or neoprene cups on the bottom of the legs, but this information has yet to reach some teaching institutions.

S.6 JEWISH COMMUNITIES

As mentioned initially, there are several branches of Judaism. Judaism is a combination of religion and ethnicity. Ethnically one must distinguish between the western Ashkenazi tradition and that of the Mediterranean Sephardi. Today, most of the world's Jewish population defines itself as Ashkenazi and, in the United States, the majority of Jewish congregations will follow Ashkenazi practice. The designs discussed so far are common to both traditions.

In the Ashkenazi custom, preliminary parts of the service frequently have each congregant praying individually although some prayers may be performed communally. A separate choir is rare, but traditionally there have been helpers who have assisted the chazzan in the song. Approximately 20% of the liturgy is chanted by the congregation and 35% is a solo by the chazzan. The remaining 45% will be silent and/or separately recited in an undertone by each member. There will be no musical instruments and no sound reinforcement except for the occasional use of the Shofar as mentioned earlier (see Reference S.1). Sephardic practice involves more unison singing by the congregation. Parts of the preliminary service are passed from one congregant to the next and nearly the entire service is sung.

During periods of solo singing, there are frequent one to three word responses by the entire congregation beside the usual Amen (see Reference S.4).

Note that a typical, observant Jewish family in all communities continues the Sabbath and holiday rituals in the home with much of the mealtime devoted to spirited singing at the table.

S.7 CHASSIDISM

The Chassidic movements had their start in Eastern Europe in the 18th century. Many of their practices are borrowed from the Sephardim. Their prayers are noted for inclusiveness, frequently embodying both Ashkenazi and Sephardic elements in the same service. In the Chassidic movements, the synagogue ranges in size from the small shtiebel to assembly halls such as the Belz World Center in Jerusalem shown in Figure S.12a. The main sanctuary there seats 6000 worshippers and is used mainly for Sabbath services (see Reference S.9).

(a)

(b)

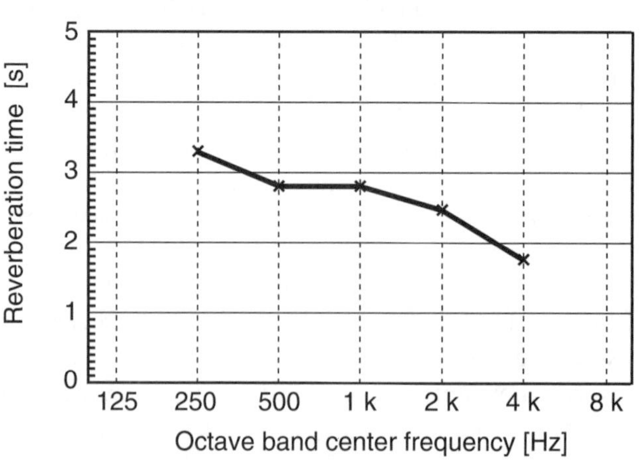

Figure S.12 (a) Belz Beis Ha-Medrash Ha-Gadol. The Beltz Chassidic synagogue that seats 6000 congregants is the largest synagogue in Jerusalem (see Reference S.9) (photo: Mendel Kleiner). (b) Reverberation time ($RT_{-5,-25}$) for the synagogue.

Despite low background noise, it needs Metropolitan Opera House-quality voices to ensure intelligibility because sound reinforcement is not allowed. The reverberation time is fairly short for a room of this size (see Figure S.12b). The Chabad Lubavich Hassidic movement in Brooklyn, New York, also has an assembly hall of about the same size.

S.8 REFORM MOVEMENT

Some decades after the emergence of the Chassidic movements of Eastern Europe, the Jews of Western Europe began to be released from the ghettos thanks to the French Revolution. The *Haskalah* was a movement among European Jews in the late-18th century that wanted the Jews to integrate into European society. One result was the Reform movement. Its founding rabbis in Germany wanted to separate ritual from moral. They immediately came into conflict with the orthodox rabbis who considered the development a threat to the traditional Jewish life. The Reform movement was brought to the United States in the mid-19th century and today claims to represent most American Jews.

The Reform movement modeled its religious services on those of Protestant Christianity. The service was shortened and many Hebrew texts were eliminated or replaced by new prayers in the vernacular. Thus services offer music as an accompaniment to prayer, as well as for other purposes, and organs are accepted in its synagogues or temples (see Reference S.4). The main emphasis is put on the Friday evening service.

Figure S.13 shows the generic floor plan of a Reform synagogue. Men and women sit together and take part in the service on equal terms. Classic Reform is

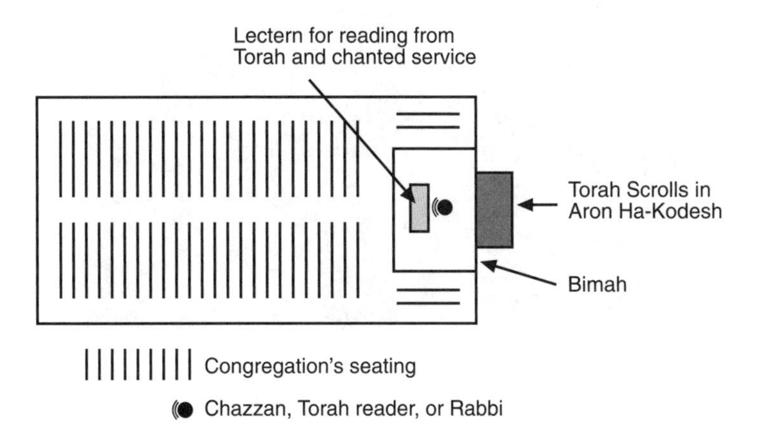

Figure S.13 Typical reform or conservative synagogue plan.

characterized by great decorum. The order of prayer is fixed. The congregation is passive, responding with Amens or something similar. There is usually little or no congregational singing, although there may be a paid choir. Those Reform synagogues that have high-quality pipe organs face the speech vs. music issue which is also faced by mainstream Protestant and Catholic churches. In Reform worship, there is no conflict with Sabbath rulings, and both musical instruments and electronics can be used as appropriate.

Another major difference between Orthodox and Reform prayer is that, in Reform prayer, the chazzan—here called a cantor and usually a professional—faces the congregation. This is, of course, a necessity caused by acoustic considerations because the cantor in both conservative and reform congregations is positioned close to the Ark at the front end of the synagogue as shown in Figures S.14 and S.15. The Reform style synagogue in Gothenburg, Sweden, is atypical since there the cantor's lectern faces the rabbi pulpit (see Reference S.10). The rabbi's pulpit, in this case, acts as a reflector for the cantor's voice because in this synagogue the cantor chants while turned away from the worshippers.

Reform synagogues may well be designed to have fairly long reverberation times suitable for the organ and other music. The sanctuary can then be optimized for high speech intelligibility using a sound system. If the reverberation cannot be attained by volume, then a reverberation enhancement system can be used. One of the largest Reform temples is Manhattan's Central Synagogue

Figure S.14 In the center, the chazzan's lectern in the reform synagogue in Gothenburg, Sweden. Rabbi's high pulpit to the left acts as a reflector for the chazzan's voice (photo: Mendel Kleiner).

Figure S.15 The Central Synagogue New York, NY, uses a LARES reverberation enhancement system. Note loudspeakers on the pillars (acoustics and sound system: SIA Acoustics Photo by Central Synagogue).

shown in Figure S.15. It is a recently renovated Reform synagogue, built in 1872, that seats more than 1000. It features a large pipe organ and a LARES reverberation enhancement system (see Chapter 7). There are a number of relatively small loudspeakers throughout the temple. The 48 loudspeakers are each fed by a suitable digital delay line and the LARES system.

A modern Reform service is characterized by flexibility. There may be lay as well as rabbinic leading of the service. There may well be a different service every Shabbat. Congregational singing is encouraged. A paid choir is optional but generally there will be a volunteer mixed choir. Some modern Reform synagogues have flexible seating and a different arrangement each Shabbat. For music there may be synthesizers, guitars, or drums, for example. Others are optimized for music with speech intelligibility enhanced by the aid of sound reinforcement.

S.9 CONSERVATIVE MOVEMENT

A reaction to the Reform movement came in the second part of the 19th century. The Conservative movement retained most of the Orthodox tradition but eliminated references to the Jerusalem temple, shortened the prayer service, but retained the Reform movement's synagogue architecture. An example of such a synagogue can be found in Albany's Temple Israel shown in Figure S.16 (see Reference S.12). Note that the cantor is again facing the congregation. Figure S.17 is another example of a large Conservative synagogue in Southfield, Michigan, the Sha'arey Zedek (see Reference S.13).

Figure S.16 The chazzan faces the congregation at the Temple Israel conservative synagogue, Albany, NY (see reference S.12) (photo: Bob Neudel).

Figure S.17 Congregation Sha'arey Zedek, Southfield, MI. The loudspeakers are located above the Ark. The lower side walls are movable, sound-isolating partitions that open into social halls used for Holiday overflow seating. Designed by Percival Goodman and Albert Kahn (see Reference S.13) (photo: David L. Klepper).

S.10 NOISE ISSUES

Conventional noise problems due to heating, ventilation, and air conditioning are, of course, common to the synagogue as well as to other places of worship. The main Conservative or Orthodox Jewish worship on the Sabbath is continuous for about three hours (Reform services are usually shorter). This means that occasionally there will be the movement of people in and out of the sanctuary, people will at times converse among themselves, and children might stir. Occasionally, the background noise level in an Orthodox Jewish worship space can become quite noticeable, but the ruling is for no noise to interrupt devout prayer. Orthodox synagogues need low noise because there cannot be any conventional electronic amplification of sound on the Sabbath. In Reform synagogues there is great decorum, and background noise levels should be set low. The desire to perform music in Reform synagogues requires low noise levels in view of their size. Implementation of low noise levels is discussed in Chapters 8 and 9.

S.11 OVERFLOW SEATING

Many synagogues, of all traditions, will have a social hall separated from the sanctuary by a folding partition. The partition can be folded so that the social hall can function as an extra space for congregants on the High Holy days of Rosh Hashanah and Yom Kippur. Such partitions are not ideal noise barriers, and people in the social hall generally do not realize that they can be heard in the sanctuary. Therefore, the partition needs to have high sound transmission loss. This matter is discussed in Chapter 9. Of course, the social hall needs to have extra sound reinforcement. The staircases needed in the rectangular style synagogues that have balconies are a problem to older people, and the noise from the footsteps on the stairs can be a problem as well.

S.12 ROOM FINISHES

In Israel, the modern trend is to make Orthodox and Conservative worship spaces to serve also as libraries and study halls so that with reasonable ceiling heights, hard ceilings, and walls lined with books, the room acoustics generally are close to optimum. As noted, the screech of chairs' metal legs on slate or tile floors may be a severe problem but one that can be resolved with rubber cups for the leg bottoms. The use of acoustic tile ceilings in many of these spaces shows a lack of knowledge by architects.

S.13 USE OF SOUND AMPLIFICATION

Traditional Jewish worship spaces have few problems regarding speech intelligibility; however, when sanctuaries become more than 15 m square there will be a lack of speaker power unless a sound system is used. Typically, sound systems for synagogues are fixed—either central or distributed. Variations of the central theme include the use of vertical column horn loudspeakers or a central line loudspeaker over the Ark. Many synagogues have an open space over the Ark, therefore, there is space in which to hide a central system such as was done in Sha'arey Zedek (see Figure S.17) (see References S.13 and S.14). Steerable pattern line-arrays, the most recent development in sound system technology, is presenting the possibility of new and superior solutions to sound amplification in Jewish worship spaces, supplementing the past solutions of central, directional, horn-based systems. These steerable line-arrays can be architecturally integrated in an optimal manner (see Chapter 10).

Another matter to note is that the prohibition toward using electric equipment on the Sabbath and many holidays prohibits the use of conventional sound-reinforcement systems. Approved sound systems for Orthodox congregations include fluidic-air-only and single-sided electronic systems that do not involve

Figure S.18 Main synagogue Shearith Israel, New York, NY is North America's oldest Jewish congregation and the first to employ electronic sound reinforcement on the Sabbath. Uses Zomet Institute (Israel) approved sound system (see Reference S.15) (photo: Dan Clayton, Clayton Acoustics Group).

zero-crossing of the electrical signal. The latter systems can provide speech enhancement with all the qualities of conventional systems.

Congregation Shearith Israel, New York, New York, (see Figure S.18) is North America's oldest Jewish congregation and was the first Orthodox congregation to employ electronic sound reinforcement on the Sabbath (see Reference S.15). The sound-reinforcement system is approved by the Zomet Institute, Alon Shvut, Israel. The Zomet Institute is an organization under the Chief Israeli Sephardic Rabbi and provided the guidelines for the Shabbat speech-reinforcement sound system.

The Hampton Synagogue, Westhampton Beach, New York, is an example of a modern Orthodox synagogue featuring the installation of the Zomet Institute approved sound system (see Figure S.19). It uses an Intellivox steerable line-array loudspeaker system to direct sound properly to the congregants without causing reverberation.

One should also note that in most Jewish congregations, certainly all Orthodox ones, it is difficult to obtain acoustic data representative of the fully seated congregation. In practice, the acoustician has only two occasions on which to obtain such data, the minor but popular festivals of Purim and Hanukkah.

Figure S.19 The Hampton Synagogue, Westhampton Beach, NY, is a modern Orthodox congregation that employs the Zomet Institute (Israel) approved sound system. Note the low, informal mechitzah separating the men's and women's seating area (photo: Clayton Group).

S.14 SUMMARY

Except for the location of the desk, from which the Torah is read, in the midst of the congregation in nearly all Orthodox and some Conservative synagogues, there is not a great difference between synagogue and church architecture and acoustics. The importance of the congregation as a sound source rather than merely listeners (except in classical Reform synagogues) distinguishes them from auditorium, theater, and concert hall acoustics. And, there remains the ban on musical instruments and sound systems in Sabbath and Holy-Day worship in many Orthodox synagogues, although air-only and single-sided electronic systems have been approved in specific cases. Additionally, musical instruments are brought into some Orthodox synagogues for family celebrations and such particular holidays as Purim, Hanukkah, and The New Month (Rosh Chodesh). In Reform synagogues there is no conflict with Sabbath rulings and reverberation enhancement can be used freely. The modern techniques of radio-frequency microphones and steerable line-arrays can make sound reinforcement effective, natural sounding, and visually unnoticeable.

Noise control remains important in synagogues and in churches. In this respect, for both churches and synagogues, and possibly mosques as well, we view some past criteria as too lenient. When the money is available, certainly the best acoustics are achieved when the same quiet of a fine concert hall also is obtained in a worship space and *RC*-15/*NC*-15 is a realistic goal. Again, each case must be weighed with reference to the overall budget. Unless noise problems are appropriately addressed, it will be difficult for the congregants to attain the desired experience.

CHURCHES

C.1 INTRODUCTION

Christian worship encompasses a wide range of music, from the timeless chants of the early Church to synthesized instruments amplified through multiple-channel loudspeaker systems. In addition, speech must be intelligible, for it communicates the divine Word to the human heart in proclamation and preaching. Because intelligible speech and various genres of music call for varying amounts of reverberation (among other acoustical parameters), acoustical and sound-system design (and their interplay) can present formidable challenges in Christian worship spaces.

One should therefore understand different liturgical (worship) approaches along with their musical characteristics. For example, some genres of music (like contemporary amplified bands or Orthodox liturgical chants) typically require acoustical environments that happen to be favorable for speech intelligibility; other genres (e.g., pipe-organ repertoire, Anglican choral literature, and Gregorian chant) sound best in acoustical spaces that—without a well-designed sound system—can make speech unintelligible even at short distances.

To provide a theological and historical context for acoustical design, this chapter first surveys briefly the development of Christian worship and its predominant elements. The spoken and musical characteristics are then translated into acoustical guidelines for the interior space and, if necessary, the sound system. Finally, some examples are given for various church environments.

C.2 HISTORICAL SURVEY OF CHRISTIAN LITURGY

During the apostolic period (approximately the first century until the death of Christ's apostles), the notion of assembly (or gathering) was essential for the early Christian Church. In the biblical New Testament, the book of Acts (2:42 ff) recounts that early Christians "devoted themselves to the apostles' teaching and fellowship, to the breaking of bread and the prayers." These four realities (apostolic teaching, communal ownership, breaking of bread, and prayers) mark the nascent structure of Christian worship, as is also exemplified by the Apostle Paul's first letter to the Corinthians (11:23 ff).

After the apostles' deaths, during the second and third centuries, Christian communities survived somewhat precariously due to persecution by the Roman Empire (see Reference C.1). Christian worship was, therefore, celebrated in small gatherings, sometimes secretly. Thus, many worship buildings were correspondingly small *house-churches*; one mid-3rd century house-church in eastern Syria had several rooms, and one of them on the ground floor held 30 to 50 people (see Reference C.2). A few large buildings were built during a period of peace after 260 but then destroyed in 303 during the Diocletian persecution; Eusebius of Caesarea describes assemblies with thousands of people, edifices with "glorious concourses," and "spacious churches" (see Reference C.3). Most early Christian worship, however, was characterized by small gatherings (see Reference C.4).

St. Justin's First Apology (ca. 150) provides the oldest description of the Eucharistic celebration, including the Liturgy of the Word—readings and preaching, prayers of intercession—and the Liturgy of the Eucharist that includes the kiss of peace, the preparation of the bread and wine, prayers of consecratory thanksgiving (in Greek, eucharistian), and Communion (see Reference C.5). Although today the readings are Biblical excerpts, in the early Church the readings included traditional teachings of the apostles because the definitive list (canon) of the New Testament books was not yet established by the Church until the formation of councils in the 4th and 5th centuries. In any case, the basic liturgical elements of *Word* and *Eucharist* are still evident today.

These initial liturgical developments began to blossom after the emperor Constantine legalized Christianity in 313 (see Reference C.6). By 604, the essentials of the rites that had progressively developed in liturgical prayer-books (sacramentaries) were compiled and codified by Pope Gregory the Great. The classical structure of the Western liturgy finally began to take shape and, due to the larger assemblies, correspondingly larger buildings for worship were built.

During this period (from the 5th century onward), the cultures, languages, and liturgies of the East (centered in Constantinople) and the West (centered in Rome) began to diverge. Centuries later, the East and West suffered the Great Schism of 1054, and their liturgies continued to develop separately. On the one hand, the Western liturgy underwent a hybridization over several centuries. In the late-8th century, Charlemagne distributed Roman sacramentaries throughout the empire, and Carolingian churches mixed in their own traditions, producing *Romano-Frankish* rituals (including Gallican rites). Approximately 300 years later, these *hybridized* Frankish rituals were adopted by the Roman Church during the reform of Gregory VII (1073-1085) (see Reference C.7). The Eastern Orthodox liturgy, on the other hand, developed more uniformly and, today, uses primarily the Liturgy of St. John Chrysostom (along with the Liturgy of St. Basil at various times during the year).

Although the Roman Catholic and Orthodox liturgies have some differences in their texts and sequence, their essence and structure (e.g., of Word and Eucharist) today are still similar. The Roman Catholic liturgy is composed of the introductory rites (including the entrance chant), the Liturgy of the Word (readings, chants, homily, short reflective silence, profession of faith, and prayers of petition), the Liturgy of the Eucharist (preparation of the gifts and the Eucharistic prayers in which the bread and wine are believed to become, in substance, the Body and Blood of Christ that are shared at Communion), and the Concluding Rites (announcements, the priest's blessing, and dismissal). The Orthodox Liturgy of St. John Chrysostom shares a similar structure of introductory rites (e.g., the Great Litany and the Antiphons, the Entrance, and the Trisagion Hymn), the Liturgy of the Word (readings, homily, litanies, prayer and dismissal of catechumens, prayers of petition, and profession of faith), and the Liturgy of the Eucharist (the Holy Anaphora prayers and Communion), and the dismissal.

Returning to the historical survey, further separations (and liturgical differences) in the West developed in the 16th century with the Protestant Reformation, involving Luther, Calvin and Zwingli, the Anabaptist movement, and Thomas Cranmer and King Henry VIII; these separations developed over the centuries into today's various Protestant denominations (e.g., Baptist, Congregational, Episcopal, Lutheran, Methodist, Moravian, Quaker, Presbyterian, Reformed, Church of Christ), and others such as the Church of Jesus Christ of Latter Day Saints (Mormons), Christian Science, and non-denominational evangelical/Pentecostal communities (see Reference C.8).

Expressing their respective theologies in worship, some Protestant denominations changed their liturgy slightly and others changed them dramatically. The Anglican Holy Eucharist, for example, remained similar to the Roman Catholic rite; Lutherans also considered their Biblically rooted worship to be liturgically faithful to early Christianity with a similar structure of the Service of the Word and the Service of the Sacrament. In the musical realm, both Anglicans and Lutherans produced new liturgical music prolifically. On the other hand, Reformed Protestant communities, following Zwingli, generally removed any sense of ritualistic ceremony and sacramental references. Zwingli himself, although a proficient musician, excluded music from worship. As another example, Calvinist Presbyterians sang psalms, although usually in unison (i.e., without harmony) and without instrumental accompaniment. Anabaptists, who desired a simpler life indicative of primitive apostolic Christianity, reflected their theology in simpler worship services in which preaching was central; musically, their services used hymns without a choir (see Reference C.9). Mormon worship is similar to some Protestant denominations with hymns, readings, and sermons, although Mormon worship spaces have no religious symbols. Pentecostal charismatic worship has significant preaching, increasingly mixing in *call-and-response* congregational dialogue and amplified music toward the culminating conclusion of the service. Most recently the late-20th-century emergence of evangelical churches established contemporary pop/rock music (often with large-scale audio-video-lighting productions) and preaching as foundational to their approach to worship and evangelization.

C.3 ACOUSTICS FOR CHRISTIAN WORSHIP

From an acoustical perspective, the various Christian liturgies are similar in that they all contain some form of speech, for example Scriptural proclamation and preaching. Some services also include spontaneous personal testimonies by members of the congregation. There naturally may be variations in microphone techniques for different preaching styles. A sermon from the pulpit would use a fixed microphone, an informal lesson from the sanctuary steps would use a wireless lapel (or over-the-ear) microphone, and a revival-style message given to a large gathering could incorporate congregational participation with a hand-held wireless microphone. Recent developments in wireless microphone technology with the ability to track the microphones using various receivers and appropriate sound routing to the various loudspeakers might work well for the Eastern Orthodox churches, which often suffer from the lack of intelligibility of the priests' chants during processions. On the other hand, in some small worship services, there may be no need for amplification at all. Even with these variations, the common need for speech intelligibility provides one general *bound* for the reverberation time (*RT*) and sound system of a given space. Typical recommended reverberation times at 500 Hz are shown in Figure 7.13 in Chapter 7 (ideally for speech $RT \geq 1$ sec).

However, the wide range of musical genres in Christian worship can make acoustical analysis difficult. For example, Eastern Orthodox liturgy generally uses the human voice and does not use musical instruments; the liturgical prayers are normally chanted by the priest and the choir, and speech intelligibility that would call for lower reverberation times is necessary (*RT* ca. 1.0 to 1.3 sec). Nevertheless, some Orthodox churches are large with reflective surfaces, thus producing significant reverberance. Furthermore, from a purely musical perspective, some chant (e.g., Russian, Greek, or Serbian Orthodox) can sound fitting in significantly longer reverberation (*RT* ca. 2 to 3 sec). Some Protestant liturgies may begin with a substantial organ prelude (e.g., *RT* 3 to 4 sec, depending on room volume) and intersperse their service with sung congregational hymns, spoken or sung (chanted) psalms or responses, and choral anthems (*RT* ca. 1 to 2 sec). Nondenominational evangelical churches might begin with an amplified band to inspire the congregation to praise and pray, followed by extensive preaching on Biblical themes; this would require shorter reverberation times and, thus, greater acoustical absorption (e.g., *RT* ca. 1 sec, although this value may be practically infeasible and somewhat relaxed for large spaces with low reverberant levels).

Besides variations in music, the size and types of worship spaces can also vary dramatically. Some churches can be intentionally small to engender and maintain a closer sense of community as illustrated by a 100-seat church in Paterson, New Jersey, with a volume of 864 m^3 (see Figure C.1). Other churches might be constructed to hold thousands of pilgrims such as St. Peter's Basilica in Rome, Italy, with a volume of 480,000 m^3 (see Figure C.2).

Figure C.1 Paterson Church of God Prophecy, a 100-seat church with a volume of 864 m^3 in Paterson, NJ (photo: Richard W. Lee).

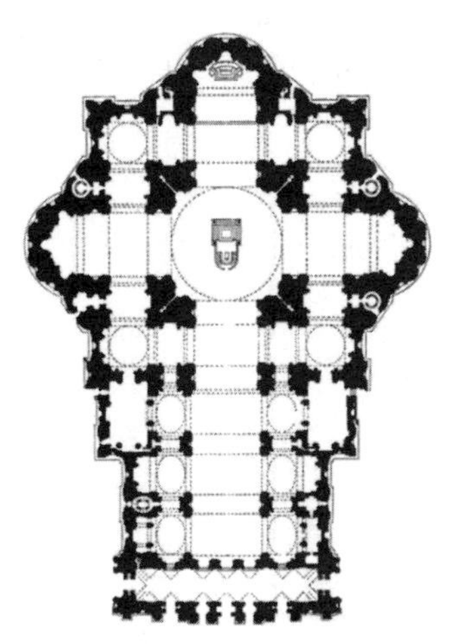

Figure C.2 Plan diagram and photograph of St. Peter's Basilica, Rome, Italy. The volume is approximately 480,000 m^3 (see References C.1, C.22, and C.23) (photo: Francesco Martellotta).

Mormon churches frequently have a connecting space separated by folding doors between their chapel and cultural hall (see Figure 9.18 in Chapter 9) that enables the two spaces to be used for separate activities; however, when the folding doors are open, it becomes one large assembly hall. Many of their churches, called *Stakes* and *Wards*, are constructed according to standard plans and adapted to local conditions. The chapel is the primary worship space and has a reverberation time of approximately 1.2 sec when occupied; pew upholstery or cushions may be used to constrain the reverberation time. All other surfaces are sound-reflecting except for the optional use of carpet. The cultural hall is multipurpose and is frequently equipped with basketball hoops. The halls use extensive perimeter ceiling sound-absorbing treatment and sound-absorbing wall panels that deflect basketballs during games. Its sound systems are discussed in Chapter 10.

The Methodist *Akron Plan* buildings of the late-19th century combined a primary worship space (a rotunda with an exposed platform and wraparound balconies) coupled to ancillary spaces such as a secondary worship annex and classrooms around the perimeter, or both (see Figure C.3). The basic plan can be modified in a variety of ways, and the room acoustics and sound system can be designed for the specific music programs and speech requirements of a particular congregation.

Instead of analyzing all types of Christian worship services (whose characteristics can overlap and sometimes differ in only minor ways), it may be clearer to discuss two distinct types that encompass most acoustical issues. Interior spaces can be characterized by several acoustical attributes such as reverberation, intimacy, strength, spectral balance, and spaciousness; these, in turn, can be described with various room-acoustics metrics (discussed in Chapter 5). In addition, some qualities are not easily reducible to numerical values such as *binaural room response* (the room's capacity to reflect early sound to the congregation from themselves

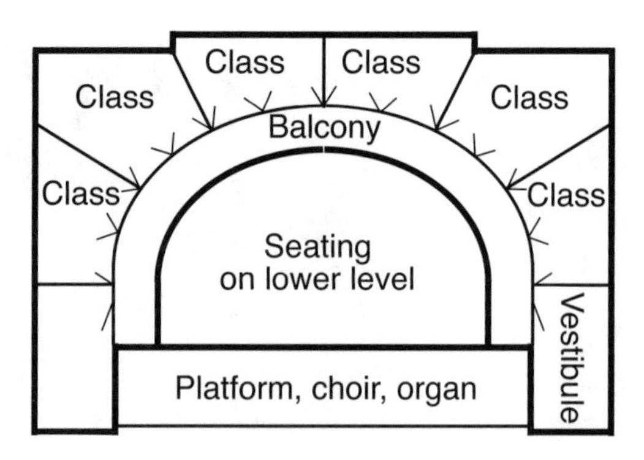

Figure C.3 In the Akron Plan, classrooms open onto a common space with a platform from which the superintendent could monitor all the classes.

and the musicians). Nevertheless, the most dominant and fundamental *acoustic dimension*, after loudness, is typically reverberation that is related to volume, average surface absorption, and—because most occupied spaces are not ideally diffuse—room geometry; thus, one may examine worship styles for which more reverberation is desired and those for which more absorption is appropriate.

Regarding reverberant spaces, an example of challenging liturgical acoustics is found in the Roman Catholic Mass that can have a wide variety of music along with speech. The Catholic Mass is frequently celebrated in reverberant spaces with acoustically low speech-intelligibility, but with desirable reverberation for instruments like organ, strings, winds, and brass. At the same time, it highly values the congregation's participation in singing/chanting and dialogical (responsorial/antiphonal) prayer.

A complementary example of nonreverberant/absorptive spaces would be large nondenominational contemporary worship spaces (including mega churches) that oftentimes use amplified pop/rock bands in more absorptive spaces in which longer reverberation is not needed and audio-video-lighting systems can compare with the most technologically sophisticated concert venues (see Reference C.10). They also include *blended* worship services that can incorporate not only contemporary bands but also orchestras and choirs—a graceful mix of the contemporary and traditional. From a reverberation perspective, these spaces remain manageably in the lower-reverberation range (e.g. $RT < 1.5$ to 2 sec); thus, their acoustical issues are less about conflicting reverberation requirements (for speech vs. music) but rather more about achieving a state-of-the-art worship venue.

Of course, not all liturgy of a given denomination requires a constant single ideal reverberation time. The need is dependent on the characteristics of the worship space, the musical style or instrumentation, and the liturgical approach; thus, a Catholic parish may prefer less reverberation whereas a Presbyterian community may prefer more.

Although reverberation is a dominant benchmark, it shouldn't diminish the importance of other key acoustical components such as intentional design for noise control and sound isolation (see Chapter 9). They contribute not only to speech intelligibility and musical dynamic range but also to more intangible elements such as the contemplative and awe-inspiring nature of a church (e.g., coupling low background noise of NC 15-25 with the natural reverberation of a large edifice) or the sense of privacy (e.g., masking noise when confessions are heard in multiple stations in an open space).

Similarly, local acoustics is important for a choir (see Chapter 7). The following comparison is a way of ordering and discussing acoustical issues with respect to a dominant and influential dimension of room acoustics, that is, reverberation.

C.4 REVERBERANT ACOUSTICS—ROMAN CATHOLIC LITURGY AS AN EXAMPLE

The Roman Catholic Mass is celebrated in a wide variety of spaces, and its liturgy is open to various types of music (sometimes even within the same Mass, that is, the international World Youth Days). Its musical history encompasses Gregorian chant, polyphonic chorales, organ works, along with *local* musical styles. Many Catholic Church documents (some specifically for the United States) provide musical and acoustical guidelines to balance universal Catholic identity with local customs.

C.4.1 Typical Liturgical Music for Reverberant Spaces

In addition to speech intelligibility, the requirements for reverberation time depend on what types of music are normal for liturgy. At the same time, the Mass offers options and a degree of flexibility in the music selection. To discern the essential musical elements of the Mass, one can review the musical and acoustical recommendations of the *Constitution on the Sacred Liturgy* (CSL, 1963), the *General Instruction of the Roman Missal* (GIRM, 2002), and STL: *Music in Divine Worship* (STL, 2007).

First, "sacred music is to be considered the more holy in proportion as it is more closely connected with the liturgical action." (CSL 12) Among many accepted genres, Gregorian chant has "pride of place" (GIRM 41), because it seamlessly "insinuates itself with the action" (see Reference C.11); moreover, it is "a living connection with our forebears . . . a bond of unity across cultures . . . and a summons to contemplative participation." (STL 72) This liturgical *bond of unity* is acoustically assisted and symbolized by an appropriate amount of room reverberation that achieves aural blending of the congregation's singing and, by its lingering nature, can raise their consciousness to contemplate things enduring and divine. Yet the Mass readily incorporates other types of music (e.g., from the local culture), "provided that they *correspond to the spirit of the liturgical action* and that they foster the *participation* of all." (GIRM 41) At the same time, worshippers from different countries at least should be able to sing the Creed and the Lord's Prayer together in Latin (GIRM 41). Finally, the responsorial Psalm (after the first Scriptural reading) should preferably be sung (GIRM 61).

For example, given that the musical selections coordinate with the themes of the Scriptural readings, a single Mass could contain the following varieties of music (see Reference C.12):

1. Organ prelude and *Hebdomada Quarta Paschae* (Gregorian chant from the Graduale Romanum)
2. Processional Chant: *Paschal Lamb, by God Appointed* (strophic hymn)

3. Psalm 110/116 (plainchant responsorial psalm)
4. Gospel Acclamation: *O Filii et Filiae* (post-Gregorian)
5. Preparation of Gifts: *We Exalt Thee* (local Gospel music)
6. Sanctus (chanted prayer)
7. Communion: *Jesu Rex Admirabilis* (short-form choral, polyphonic); Communion Processional: *Christus Resurrexit* (modern chant, folk-hybrid)
8. Music after Communion: Duruflé *Pie Jesu* (solo from large-form choral)
9. After Mass: *Christian, Do You Hear the Lord*? (sung in chant form)

Concerning instruments, the pipe organ is valued highly: "In addition to its ability to lead and sustain congregational singing, the sound of the pipe organ is most suited for solo playing of sacred music in the Liturgy at appropriate moments" (see Reference C.13). But other instruments can be used: "While the organ is to be accorded pride of place, other wind, stringed, or percussion instruments may be used . . . according to longstanding local usage, provided they are truly apt for sacred use or can be rendered apt" (see Reference C.14). Similar to the evaluation of music, *aptness* depends on how smoothly and naturally a given instrument can integrate into the liturgical action.

C.4.2 Acoustics for Liturgy

How does this translate into acoustical requirements? One place to start is to review a recent document, *Sing to the Lord*, that attempts to offer broad acoustical guidelines: "While individual ministers of the Liturgy, ensembles, and even choirs can be sound-enhanced through amplification methods, the only amplification of the singing assembly comes from the room itself. Given the primacy of the assembly's song among all musical elements of the Liturgy, the acoustical properties of the worship space are critical. For this reason, specialists in acoustics should be consulted when building or modifying liturgical space" (see Reference C.15). Moreover, if the "amplification" of the assembly "comes from the room itself," the room should not have a large proportion of absorbing surfaces (see Figure C.4 for an example of excessive surface sound absorption), and the location of and intention behind those surfaces should be carefully considered by an acoustic design professional who is sensitive to the congregation's liturgical and musical needs.

Sing to the Lord continues: "If each member of the assembly senses his or her voice joined to the entire community in a swell of collective sound, the acoustics are well suited to the purpose of a gathered community engaged in sung prayer. If, on the other hand, each person hears primarily only his or her own voice, the acoustics of the space are fundamentally deficient" (see Reference C.16). This test is an indicator of suitability, although not entirely sufficient; a person's voice

Figure C.4 An example of excessive sound-absorptive material applied to pews and floor (photo: Dennis Fleisher, Musonics).

level is adjustable, and it is proportionally nearest one's own ears and, with little effort, can be raised audibly (to one's self) above any typical reverberant level. Nevertheless, the document's point is that a responsive, reverberant room is preferred for this type of liturgy.

Sing to the Lord also cautions against sound-absorptive materials, for example, "carpet, porous ceiling tiles, soft wood, untreated soft stone, cast concrete or cinder block, and padded seating" (see Reference C.17). While recognizing what should be avoided, one should recognize the desired role of heavier (e.g., thicker), acoustically reflective materials (especially for low-frequency reflection and sound isolation) that are mounted appropriately to achieve full-frequency room reverberation and response. For example, thick pine (i.e., soft wood) interlocking planks screwed onto studs will reflect low-frequency sound more effectively than thin oak (hard wood) panels attached by clips. Also, the appropriate use of heavier constructions and materials can achieve desired acoustical warmth, that is, a favorable bass ratio above 1.0 (see Chapter 5 for values based on reverberation time).

Finally, the subcommittee writes, "The acoustics of a church or chapel should be resonant so that there is no need for excessive amplification of musical sound in order to fill the space and support the assembly's song. When the acoustics of the building naturally support sound, acoustic instruments and choirs generally need no amplification. An acoustically dead space precipitates a high cost of sound-reinforcement, even for the organ" (see Reference C.18). Despite inaccurate terminology (the acoustical term resonant normally refers to an undesirable emphasis of a certain frequency band or bands), the document makes clear a preference for a higher reverberant level (i.e., acoustical strength index/room gain; see Chapter 5) and, correspondingly, either less surface absorption or a smaller room volume, given that the desired range of reverberation time is achieved. It is not necessarily the case, however, that ideally responsive and reverberant acoustics will eliminate the need for any sound system; a tall church with long reverberation may need a sophisticated sound-reinforcement system not only for speech but also for musicians, because the natural reverberant sound level will be inadequate for large congregations. Finally, it is rare that a pipe organ (nonelectronic) in a dead space is amplified through a conventional sound-reinforcement system, although a reverberation-enhancement system offers a possible (though generally costlier) solution.

What then is an appropriate reverberation time? The sample musical varieties in the list correspond to a potential wide range of reverberation requirements for a single liturgy: from preaching and Gospel music ($RT < 1$ sec) to various forms of chant (*RT* approximately 1 to 2 sec) to organ music (*RT* about 2 to 4 sec; some churches offer more than 10 sec).

Beyond a certain point, however, more reverberation is not always better. Listening tests indicate that an upper limit of reverberation time of 3 to 4 sec is preferred. Using longer values does not improve listening conditions for music; for example, Martellotta shows the correlation between reverberation time and normalized global rating for various vocal and organ motifs. For excerpts of Gregorian chant ("Pange Lingua" in Phrygian mode), polyphonic choral music (Randall Thompson's "Alleluia"), and organ music (J. S. Bach's "Fantasia" in G minor), the listener's preference corresponded to reverberation times between 2 to 4 sec if all three motifs were to be accommodated within a single reverberant space (see Reference C. 22).

One should also consider a room's reverberant level—dependent on room volume and surface absorption—for example, if the reverberant level is low and the listeners are close to the source and also to early-reflecting surfaces, somewhat longer reverberation times can be tolerated although the problem of balancing reverberation with speech intelligibility remains.

These reverberation extremes are difficult, if not impossible, to reconcile by purely physical-acoustics means and, thus, present a choice: (1) one can design the space to accommodate speech (i.e., prescribing high surface absorption and shorter reverberation time) and utilize electroacoustic reverberation-enhancement systems for music; or (2) one can design the space to accommodate organ music (i.e., prescribing reflective materials and longer reverberation times) and, for speech intelligibility, employ well designed sound-systems that focus acoustic radiation on the congregation and avoid irradiating the room's upper reverberant volume. Often, the latter approach can be more affordable as long as other acoustical issues (echoes, sound isolation, noise, etc.) are competently addressed.

In either case, users (e.g., building committees) should always distinguish between (and not conflate) a space's natural acoustics and the sound system; moreover, the design and performance of the sound system typically depends on the already designed or existing room's acoustics unless an acoustician is brought into the design phase of a new construction.

As discussed in Chapter 10, there are various approaches to sound systems in reverberant spaces, including steerable line-arrays that can focus sound on the seating area more effectively than conventional clusters of loudspeakers. One example is found in St. Paul's Episcopal Church, Rock Creek, Washington, D.C. (see Figure C.5). Even though interior design issues made necessary a higher position than ideal for the line-arrays, they still focus the sound on the congregation while remaining visually unobtrusive.

The acoustician accomplished this by using custom arrays with downward aimed high-frequency horns in place of the normal full-range drivers at the bottom of the array. Moreover, the line-arrays radiation patterns were optimized through computer-modeled *beam-shaping* rather than conventional *beam-steering*. This specifies which areas should be irradiated—for example, pews—and which areas should not—upper walls and ceiling). The side-wall reflections from line-arrays are not destructive for speech intelligibility in a room of this relatively small size, because they arrive soon after the direct sound and are oriented downward, that is, toward the absorptive congregation and away from the reflective upper volume.

One potential issue with steerable line-arrays is that there may be a cutoff frequency (e.g., around 100 Hz) below which the manufacturer may choose to *roll off* the radiated sound with a high-pass filter. (In some models the cutoff frequency is directly related to the limited size and excursion of the driver.) This is reasonable for amplification of speech, whose spectral content—especially for intelligibility—is concentrated in higher frequency bands. Music, however, if amplified, may

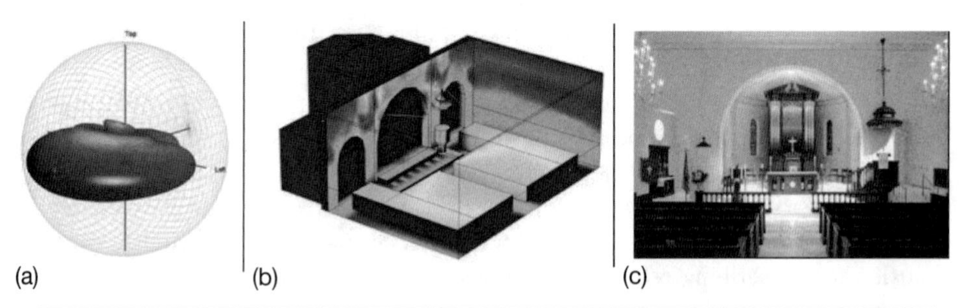

Figure C.5 Performance of steerable line-arrays. Design issues made a higher position more necessary than ideal for the line-arrays, yet they are still able to focus the sound on the congregation while remaining visually unobtrusive. (a) Figure depicts the high directivity of a steerable line-array (3-D plot from Duran Audio BV). (b) The computer plot shows the predicted irradiation by the direct sound whose level is proportionate to the brightness in the plot. (c) Corresponding to the computer plot, the photo shows an installation in St. Paul's Episcopal Church, Rock Creek Parish, Washington, D.C. (project acoustics, computer plot, and photo: Clayton Acoustics Group).

require separate loudspeakers for lower frequency content, although such *split* sound systems introduce other challenges, for example, differences in distance attenuation and directivity (between line-arrays and conventional loudspeakers), visually aesthetic integration, and ensemble hearing; moreover, if amplified music is an essential part of a church's worship, this should be communicated clearly to the acoustician.

Other approaches to sound-reinforcement include distributed systems in which loudspeakers are mounted on structural columns or even on the backs of pews (so-called pew-back systems discussed in Chapter 10). In distributed systems, the loudspeaker signals are adjusted in level and delay to approximate spatial *samples* of the sound wave that would emanate from the natural sound source.

As covered in Chapters 7 through 9, acoustical planning includes several other aspects of room acoustics, space planning, sound isolation, and mechanical system noise and vibration control. In music space planning, however, one common question regards the placement of the choir, cantor, and instrumental musicians. One issue is visual. During parts when the congregation needs to be led, the cantor should be visible; when leading is not necessary, visual distractions from the liturgy ideally should be minimized, whether from cantors, choir, or musicians who themselves are considered part of the worshiping congregation. (It may not be always feasible to implement this, especially if the organist needs to hear and direct ensemble singing from a console near the front of the church.) Figure 7.29 shows an arrangement with this purpose.

There should also be adequate sight lines and sound paths for ensemble hearing. If the organ pipes are located mainly in the upper rear of the church (where, in many churches, they efficiently excite the room's reverberance and radiate a smoother sound-level distribution on the congregation floor), there are viable reasons for having the choir either near the pipes or away from them. One reason for having them near the rear pipes, in the rear balcony, is that the choir is most closely time-aligned with the pipes. Furthermore, the singing can sound more spatially unified and blended—more unified because the perceived source width is narrower due to the long distance and more blended because the blended reverberant sound is stronger at the congregation's ears than the direct sound (due to distance attenuation). On the other hand, positioning the choir farther from the (typically louder) organ pipes and nearer the congregation helps to balance the choir's and organ's perceived sound levels at each listener (see Reference C.19), and also allows for easier intra-choral ensemble hearing, because the near-field levels of organ pipes can sound deafening. Furthermore, having the choir near the congregation reflects their membership in the gathered assembly and encourages congregational participation (see Reference C.20). In summary, there are reasons for placing the choir in either position, and their respective compromises must be weighed and negotiated by the users.

C.4.3 Example Projects: Reverberant/Semi-reverberant Spaces

The following projects are examples of acoustical sound systems for various reverberant and semi-reverberant environments.

Conventional Clusters

Conventional clusters of loudspeakers that are suspended above the sanctuary are sometimes avoided by clients and designers because they are oftentimes easily visible. Nevertheless, they can be less obtrusive in some spaces, for example, those with tall ceilings. Figure C.6 shows a central cluster system suspended approximately 25 m above the altar.

Another example in Figure C.7 illustrates how loudspeakers may be hidden visually behind a screen that is acoustically transmissive at the frequencies of interest (if not perfectly sound-transparent over the audible frequency spectrum). The central array is concealed within a suspended central element over the altar platform.

A variation on this theme is an *exploded cluster* in which high-directivity speakers are suspended in zones to bring the direct sound closer from each loudspeaker to areas of the congregation. In Figure C.8, a church with a wide nave

Figure C.6 A conventional loudspeaker cluster can be tastefully mounted so that it serves the congregation without necessarily being visually obtrusive, especially in buildings with tall heights. This photo shows a loudspeaker cluster installation at the Cathedral of the Immaculate Conception in Syracuse, NY. The cluster in the photograph is suspended from the roof about 25 m (80 ft) high above the altar (project acoustics and photo: DCi Sound).

Figure C.7 A loudspeaker cluster may also be hidden within architectural elements given acceptable sound transparency. Installation at St. Francis of Assisi Catholic Church, San Antonio, TX (project acoustics and photo: Dennis Fleisher, MuSonics).

Figure C.8 An example of an exploded cluster, suspending high-directivity loudspeakers over several zones (project acoustics and photo: DCi Sound).

and long reverberation time uses loudspeakers that have individually assigned frequency-equalization and delay that is appropriate to its coverage area. This solution has characteristics of both clustered and distributed systems.

Distributed and Pew-back

Distributed systems with simple cone loudspeakers on columns are sometimes installed, but these are not recommended by the authors. Such systems require the room to be non-reverberant (absorptive) in order to work well. In any case, a system using delayed column (array) loudspeakers would work better.

Pew-back systems are considered the equivalent of giving every listener his/her own high-quality radio. Because most sound from the small close-mouthed loudspeakers is absorbed by the worshippers' clothing and the pew cushions, this system also works well in reverberant environments. Chapter 10 features a discussion and several photos of pew-back systems.

Vertical Steerable Line-arrays

As mentioned earlier, steerable line-arrays are designed for especially reverberant environments and have aesthetic advantages because of their relative unobtrusiveness. More directive and focused than a central cluster, they emit a continuous wavefront instead of a sampled one, in the way that time-aligned (signal-delayed) distributed speaker systems attempt to do. Figure C.9 shows a steerable line-array system installed for the renovation of a cathedral in Rochester, New York. The line-arrays are slim and positioned to blend in with the long features of the stonework. Another example is shown in Figure C.10. Their unobtrusive slimness blends in with the liturgical images and action in the sanctuary. In this project, the acoustician offered a choice between two different manufacturers for the steerable line-arrays. After listening tests in situ with both models, the church staff

Figure C.9 In this photograph of Sacred Heart Cathedral, Rochester, NY, the line-arrays are installed on the side columns with the floral arrangements (project acoustics and photo: DCi Sound).

Figure C.10 In this reverberant church, Mission Basilica, San Juan Capistrano, CA, two steerable line-arrays were installed on the sides of the second set of columns (project acoustics and photo: Dennis Fleisher, MuSonics).

and rector selected one of them based on what they preferred for music. This does not necessarily mean that one manufacturer is always preferable to another, but various models may have intentionally different sound quality.

The photo of the Holy Cross Church, Dewitt, New York, in Figure C.11a shows a concealed steerable line-array to the right of the statue and mounted nearly flush with the wall. The photo of the Cathedral of Christ the King in Atlanta, Georgia, in Figure C.11b presents another example of blending in the loudspeaker by simulating the stone pattern on its surface. Although a smaller line-array would suffice for speech reinforcement alone, the users' dual need for music and speech reinforcement in their reverberant church was addressed by using full-range steerable line-array loudspeakers for the nave. Lower-resolution line-arrays were used for the transepts and sanctuary; in the transept in which contemporary musicians perform, the line-array had two beam configurations: one for the congregation in the pews and another for the musicians. An under-pew subwoofer for the transept was also installed. Figure C.11c shows the use of line-arrays in a Greek Orthodox church with two arrays mounted behind grilles above the chancel.

On the other hand, oftentimes the loudspeakers cannot be placed in acoustically or visually ideal positions. At St. Paul's Episcopal Church (Figure C.5 right), the new *wineglass* high pulpit and suspended canopy on the front wall occupied the ideal position/height for the line-arrays and, furthermore, could obstruct the sound. The

(a)

(b)

(c)

Figure C.11a (a) To the right of the statue, a well-concealed steerable line-array is mounted nearly flush with the wall. The church, Holy Cross Dewitt, was a new construction in which the room acoustics and sound system were integrated into the design (project acoustics and photo: DCi Sound). (b) Another good example of making speakers blend in, in this case by replicating the stone pattern on its surface at the Christ the King, Atlanta, GA (project acoustics and photo: Clayton Acoustics Group). (c) In St. Nektarios Greek Orthodox Church, Charlotte, NC, two steered line-arrays are hidden behind rectangular grilles above the chancel (project acoustics and photo: Design 2020).

Figure C.12 First Presbyterian Church, New Bern, NC (project acoustics and photo: Clayton Acoustics Group).

acoustician thus had to raise the position of the loudspeakers and design the radiation patterns to reach the congregation without excessively reflecting sound from the top of the pulpit's canopy, the upper side, and the rear walls.

Figure C.12 shows a Presbyterian church that received its first sound system in 2001 since having been built in 1822. The church's reverberation time (ca. 2 sec) is appropriate for its pipe organ but hinders the spoken word. The steerable line-arrays were selected by the church as the most cost effective and least intrusive solution to the historic character of the building.

Figure C.13 shows the First Church of Christ, Scientist, Greenwich, Connecticut, where a new speech-reinforcement sound system was designed to provide speech intelligibility, while not exacerbating acoustical problems from the ceiling's dome. The sound system simultaneously serves two church readers who speak from a raised platform at the front of the auditorium, plus testifiers who use hand-held wireless microphones from their pew seats throughout the auditorium. Two 2.5 m (8 ft) steerable line-array loudspeakers were surface-mounted on pilasters on the front auditorium wall and left relatively undecorated, following the congregation's preference. Other sound system features include a *handsfree* mixing system for church services and testimonial meetings, supplementary loudspeaker system for the balcony, program monitor loudspeakers for ancillary and support areas, and assisted-listening system for the hearing impaired.

Figure C.13 First Church of Christ, Scientist, Greenwich, CT, uses two 2.5 m (8 ft) steerable line-array loudspeakers surface-mounted on pilasters on the front auditorium wall (project acoustics and photo: Clayton Acoustics Group).

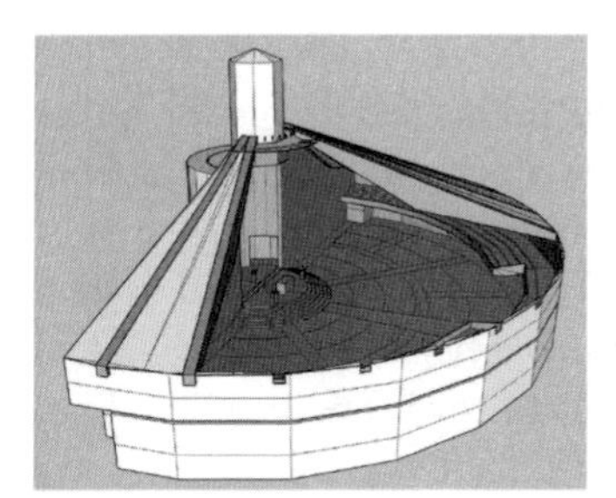

Figure C.14 Our Lady of the Rosary Catholic Church, Qatar. Note the two line-arrays on the left-hand side of the photo; a computer model of the space displays its geometric configuration (project acoustics, graphics, and photo: Clayton Acoustics Group).

Figure C.14 shows the first Christian church to be built in the Persian Gulf emirate of Qatar since pre-Islamic times: Our Lady of the Rosary Catholic Church was consecrated and opened for Easter in March 2008. Four other Christian churches—Anglican, Coptic, Greek Orthodox, and an interdenominational center for 11 Indian congregations—will be built nearby. This complex of five churches will serve Doha's growing Christian expatriate population in Qatar. The worship space plan is a 180-degree fan of 31 m radius with a flat main floor and simple wood bench seating surrounded by a shallow balcony with a 1.5 m high circular sanctuary platform and apse at the center of the long wall.

The building seats approximately 2700 people and provides standing room for another 800. The primary construction materials are plaster on solid masonry, marble and terrazzo floors on a grade slab, and relatively few windows and doors. While this is not a typical geometry for a Catholic church, it is, nevertheless, not uncommon. Reverberation time ranges from 5 to 6 sec over the audible frequency range, and the unamplified spoken word is unintelligible. Given the long front-to-back distances and the need to cover both a large half-circular main floor and a balcony, 5 m (15 ft) beam-shaping line-arrays were chosen by the acoustician. Line-arrays were mounted with their acoustic centers below the platform microphone level to reduce the potential for feedback, yet at a minimum height above the standing congregation to irradiate them with a sound-beam at a grazing-incidence angle. Small, passively steered line-array loudspeakers were used for *voice fill* into the apse area of the sanctuary.

C.5 ABSORPTIVE ACOUSTICS—EVANGELICAL/BLENDED WORSHIP AS AN EXAMPLE

Another style of worship is found in contemporary evangelical and pentacostal churches, sometimes large in scale and linked by several associated campuses (see Reference C.21); however, this style can also be blended with more traditional services and utilized in smaller buildings. Although the contemporary worship service can be structured and well organized, it does not usually appear traditionally ritualistic. If one were to compare its structure to the historic liturgical pairing of the *Liturgy of the Word* and the *Liturgy of the Eucharist*, contemporary worship emphasizes the former more by way of Bible-based preaching (sometimes with dialogical call-and-response). Oftentimes it is preceded by a musical concert offering praise and worship to God, while also preparing the congregation to hear the Scriptures receptively. In some communities a Eucharistic component (i.e., some form of breaking bread) can sometimes follow as well.

C.5.1 Typical Worship Music in Spaces with Absorptive Acoustics

In contemporary evangelical worship, amplified bands are the main source of instrumental music. This does not, however, exclude traditional instruments such as pianos and string orchestras (Figure C.16A). There may also be instruments particular to a certain culture or country. Capacities can approach those of a moderately sized sports stadium with 10,000 to 35,000 people seated; although the number is usually fewer, and sometimes is spread over several campuses (a *megachurch* is defined as having over 2000 seats).

Involving this many people in a single worship experience does, of course, require the use of modern audio/lighting/video technology: not only extensive and well-designed sound systems, but also TV cameras and large TV screens (Figure C.15). Even smaller venues, however, which would not be called megachurches, can be equipped with sophisticated sound and video systems. In this type of worship, video can play a large role, including on-screen projection of the words for congregational singing, pre-recorded and live video program material, and other applications—all of which must be coordinated, synchronized, and integrated with the sound system.

C.5.2 Acoustics for Worship in Absorptive Spaces

The physical characteristics of these churches generally include short reverberation times relative to the large room volumes with even the largest churches not exceeding 2 sec occupied, and the range 1.0 to 1.5 sec is more typical. Short reverberation times are preferred to maintain clarity of both speech and amplified music. Even if the ceiling is made sound-reflecting to assist congregational singing, sound-absorbing treatment on the rear wall and side walls, and/or above suspended sound-reflecting panels (and other visually hidden surfaces) help to achieve a short reverberation time.

There are also sloped floors with good sight-lines that can reduce reverberation by presenting a greater surface area of absorption by the congregation. The

Figure C.15 (a) Contemporary worship spaces can take on different forms: An example with piano and string orchestra, Christian Life Assembly, Camp Hill, PA (project acoustics and photo: Design 2020). (b) Contemporary worship spaces can take on different forms: An example with concert sets and lighting, The Gathering Place Worship Center, Lake Mary, FL (photo: Design 2020).

use of wide fan plans or circular or half-circular plans helps to bring the congregation as close to the stage as possible despite the high capacity. Large balconies are often included. As in unamplified concert halls, these features can decrease the reverberation by directing the sound toward the rear (due to the fan-shaped plan) and by increasing the absorption surface area (e.g., the balcony surfaces, among others). For reasons both acoustical and ergonomical, upholstered seats are commonly used that can reduce excess reverberation in the room at mid and high frequencies, depending on the thickness and porosity of the upholstery.

The stages can typically accommodate large musical groups, oftentimes sized for a full symphony orchestra. Stages frequently include lifts that allow orchestras to rise into the view of the congregation. The sound system itself can be managed by audio mixing consoles, one for the reinforcement to the congregation, one for fold-back coverage so musicians and clergy can hear themselves, and one for transmission to radio and TV broadcasting stations and for coverage of lobbies, rehearsal rooms, classrooms, and offices that are part of the building complex.

Contemporary worship has had some impact on mainstream Protestant and Catholic churches and has provided some both a traditional liturgy and a contemporary worship experience. Although some have made the investment and installed additional sound equipment, most prefer to supplement the fixed speech-reinforcement system with portable equipment that can be used for other events such as outdoor services.

C.5.3 Example Projects: Absorptive/Semi-absorptive Spaces

The Calvary Assembly of God's main hall for prayer in Orlando, Florida is designed like a large modern theater and is a typical design (see Figure C.16). It seats 3000, and all surfaces except the balcony rear wall and the main floor rear wall are sound reflecting. The volume-to-seating ratio provides a reverberation time of 1.4 sec when fully occupied. The control console for sound-reinforcement is in an ideal location in the center of the balcony. The worship hall has two independent sound systems, either one capable of providing coverage and intelligibility for every seat. One is a broad-front central loudspeaker system with center, left, and right loudspeakers, and the other is the comprehensive ceiling and underbalcony-soffit distributed system. The impression of envelopment by the spoken word and the reinforced music is enhanced when both systems are used.

Another example of a powerfully amplified space belongs to Faith Alive Ministries in Chesapeake, Virginia shown in Figure C.17. The 2200-seat facility houses a fan-shaped seating area with theater seats. A significant acoustical feature is that the entire ceiling acts as a bass-frequency absorber to control low-frequency room response. Other absorption measures include fabric-wrapped architectural clouds (overhead panels) with heavy sound-absorbing fiberglass behind them.

Figure C.16 Calvary Assembly of God, Winter Park, Orlando, FL, 3000 seats, contemporary worship, use of audiovisual effects and orchestra (architects: Schweizer Associates, acoustics & sound-reinforcement: Klepper, Marshall, King Associates, photo: from the Church).

Figure C.17 The church of Faith Alive Ministries is an example of integration of audio, video, and lighting for contemporary worship services (project acoustics and photos: Design 2020). The photo on the left is of the line-array systems that are similar to those used in concert venues. The photo on the right shows the sound-console and the interior space.

The room's side walls are sculpted to minimize flutter echoes and reflect sound for audience support. The large line-array audio system is capable of high sound levels; this type of line-array loudspeaker is also acoustically directive but is not intended to be as focusing as the steerable line-arrays discussed earlier for reverberant environments.

A contrasting example, Christ Church in Columbia, South Carolina, is shown in Figure C.18. This church purchased an old metal warehouse building in downtown Columbia and renovated it into a more traditional room, fitting also for small chamber ensembles and classical music. The acousticians developed the room response to have an even reverberation character, and the ceiling *clouds* provided congregational support. The volume above and slots between clouds add low-frequency absorption. A gypsum isolation barrier was installed under the roof deck to minimize intrusion from rain noise. For sound-reinforcement, a distributed speech system was implemented.

Another interesting project, the Westover Church, Greensboro, North Carolina, shown in Figure C.19, combines the traditional and the contemporary. The room seats approximately 2500 and is designed to have a performing-arts stage. The acoustical design utilizes micro-perforated ceiling clouds for absorption, resulting in short reverberation times. For more traditional sources (e.g., choir, orchestra, etc.) and classical concerts, an installed reverberation enhancement system is used. The space is regularly employed for classical concerts, including hosting the Greensboro Symphony's pops concerts.

The dual-purpose First Baptist Church in Simpsonville, South Carolina, hosts both traditional and contemporary services in the sanctuary. Thus, the acoustical design had to cover a variety of uses from choir and organ to a full contemporary

Figure C.18 This venue, Christ Church, Columbia, SC, shows how a warehouse can be transformed into a space for worship and small chamber-music ensembles (project acoustics and photo: Design 2020).

Figure C.19 These photos of the Westover Church, Greensboro, NC, illustrate a venue that includes a reverberation enhancement system so that the space can accommodate both traditional and contemporary liturgy and music (project acoustics and photos: Design 2020).

Figure C.20 These photos show the Simpsonville First Baptist Church, Simpsonville, SC. The fold-back loudspeakers on stage provide monitoring for the musicians (project acoustics and photos: Design 2020).

worship band. Support was needed for congregational worship as well as natural acoustical support of the building's electronic organ system. The HVAC noise was made low, which increases dynamic range and makes preaching and music possible at softer, dramatic levels. The clouds over the audience support the congregational worship. The sweeping balcony also supports the congregation in the lower level. The under-balcony ceiling was faceted for sound scattering in that area, and absorption was placed judiciously to dampen potential flutter echoes between parallel surfaces. The ceiling space behind the clouds contains catwalks and provides low-frequency absorption. The space has a hybrid organ (or *combination* organ with both windblown pipes and digital voices amplified through loudspeakers) with front and antiphonal chambers and additional loudspeakers throughout the cat walk for the organ. A large canopy over the choir helps with stage support. The church interior is shown in Figure C.20.

The audio system in the space is distributed among the two major theatrical catwalks and hidden by a stretched fabric scrim. Large arrays of loudspeakers cover the main seating areas; areas over and under the balcony, as well as the front rows, are covered locally. The wide-format video system is centered around engineered rear-screen projection surfaces for the audience and a front-projected *confidence monitor* above the balcony. A complete theatrical lighting system illuminates both the traditional and contemporary services from two major catwalks plus over-stage positions.

Elevation Church consists of 4000 to 5000 people across several campuses in Charlotte, North Carolina. Figure C.21 shows the broadcast/central campus where the sermons are filmed on Saturday night and transmitted to its other video venues for Sunday. It has a low ceiling in an existing commercial warehouse. To reduce noise from the space into the neighboring buildings, the acousticians cut the slab, roof, and building structure to minimize sound transmission; they also used heavy walls for further sound isolation. There are bass-frequency absorbers (*bass-traps*) built into the side and back walls to maintain an even room response. The distributed audio system includes significant power for bass/subwoofer loudspeakers.

Finally, the Willow Creek Community Church, South Barrington, Illinois, (7200 seats) shown in Figure C.22 uses a stage house that includes a fly tower. The windows flanking the stage present a large reflective surface that the acousticians addressed by splaying them wide from the stage and installing retractable acoustical banners (offering absorption) that can be lowered over the windows. The floor beneath the seating areas was designed to be reflective to support congregational participation. Low-frequency (sub-bass) speakers are mounted under the platform and in the speaker cluster area, and front fill speakers cover the front rows. The acoustical planning also included a large green room, rehearsal rooms for small

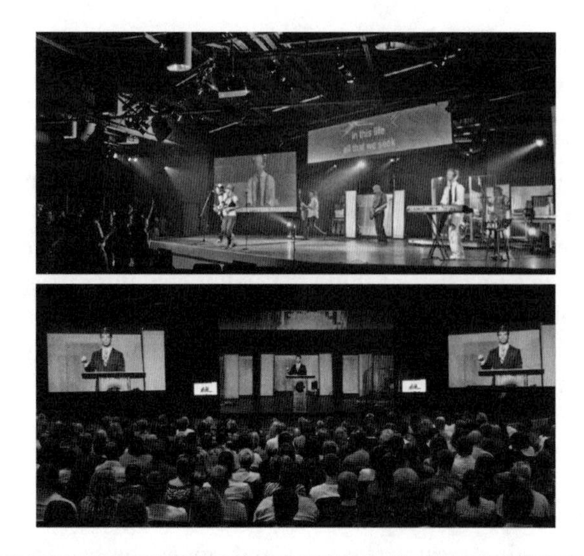

Figure C.21 Elevation Church consists of 4000 to 5000 people across several campuses in Charlotte, NC. The room shown is the central campus where the sermons are filmed on Saturday night and transmitted to the other campuses for Sunday worship. As shown in the upper photograph, song lyrics are shown on the screen to encourage congregational participation (project acoustics and photos: Design 2020).

Figure C.22 The auditorium of the Willow Creek Community Church in South Barrington, IL (project acoustics and photo: Acoustic Dimensions).

ensembles, instrumentalists, vocals, orchestra, drama, and dance among many other areas.

C.6 SUMMARY

Christian worship spaces can house a variety of musical genres and worship styles. Some liturgies, moreover, use programming with conflicting acoustical goals, for example, speech intelligibility and longer reverberation. As discussed in this chapter and in Part I of this book, the variety of acoustical demands of Christian worship spaces should be addressed by a competent acoustician:

1. Room acoustics to determine room shape, size, and materials
2. Music space planning to ensure ensemble hearing and sight lines, and to discuss locations for practice areas and for choirs and musicians
3. Sound isolation to prevent unwanted sound in important listening and worship areas
4. Mechanical system noise and vibration control to minimize the distraction from HVAC noise and, in concert with sound isolation, to provide a wider dynamic range
5. Sound-reinforcement systems to optimize the natural acoustical environment according to the needs of the worship service and to allow the combination of the natural acoustics and the sound system to meet both speech and music requirements.

"Faith comes from what is heard" (Romans 10:17); moreover, by integrating these complex but essential acoustical factors from the earliest phases of design, the multiple demands of these spaces will best serve those who worship in them.

MOSQUES

M.1 HISTORIC DEVELOPMENT

Islam's history suggests that Muhammad and his followers probably prayed outdoors on their journeys and in tents before the first mosque building was constructed. This was a rectangular building, 4 m high, 26 m front-to-back, and 30 m wide. It was constructed of mud bricks with a date palm-supported roof. This rectangular form has remained the most popular and traditional of mosque architectural forms. It was built in 622 CE 5.5 km southeast of Madinah during Muhammad's journey from Makkah to Madinah, Saudi Arabia. In 633 CE, Muhammad built a house, courtyard, and mosque in Madinah that was also rectangular and made of mud bricks (see Figure M.1). This was a courtyard mosque mostly open to the air with a roof over a portion of the area, and this rectangular portion was the main praying area. The enclosed portion of this mosque along with the previous mosque became the prototypes (see References M.1, M.3).

Rectangular floor plans were used for the earliest fixed mosques. The Caliph of Egypt built the Al Azhar mosque in Cairo in 970 CE introducing arches and a dome. For a long time, this mosque and its associate school was the center for Sunni Islam theology (see Reference M.5). The Hagia Sophia was built to serve as a church in 535 CE, but after the Turks took control of Istanbul (formerly Constantinople) and converted the church to a mosque, it became a model for other mosques of varying sizes, undoubtedly influencing the design of the Cairo Al Azhar mosque (see Figure M.2). Instead of rectangles, shapes such as circles, hexagons, and octagons became popular because of a strong Turkish influence.

The Safavids, who extended the influence of Iran (Persia), established the general provision of a school (madrassa), mausoleum (turba), bathhouse (hamam),

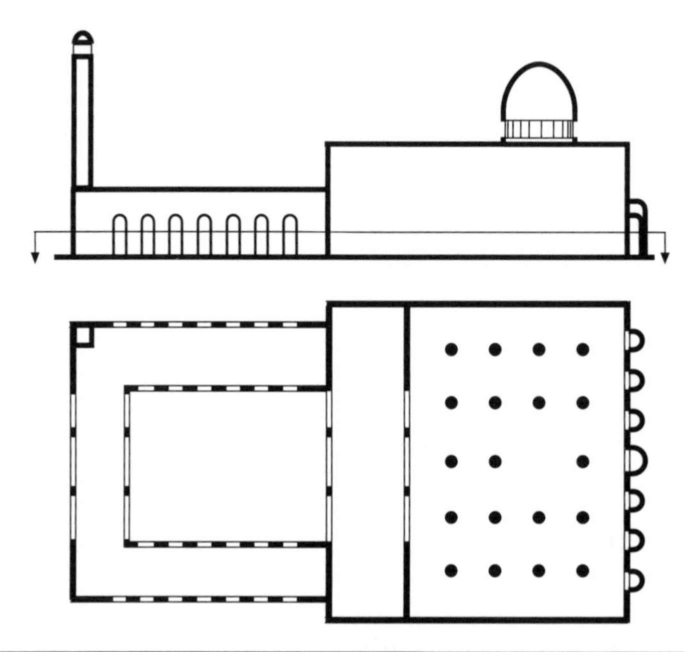

Figure M.1 Traditional Mosque architecture. Note dome, minaret tower, and enclosed rectangular courtyard.

Figure M.2 Hagia Sophia Mosque interior (photo: J.-H. Rindel).

library, inn (khan), hospital, and shops to surround and coordinate with the mosque (see Reference M.5). However, most modern mosques are built on some variation of the basic rectangular form established by Muhammad, but they can vary widely in size (see References M.1, M.2, M.3).

M.2 WORSHIP CHARACTERISTICS

Islamic worship takes two basic forms: individual or group prayer and listening to the clergyman's (imam) sermon (khutbah). With group prayers, the worshippers are located in rows parallel to the quiblah about 1.2 m apart as shown in Figure M.3.

Standing, kneeling, sitting, and prostrating oneself are all part of standard worship (see Reference M.2). Normally, the imam is not located on the minbar, but rather chants, sits, stands, and kneels the same as the congregation. However, some modes of prayer have the imam chanting solo and the congregation responding (see Reference M.6).

A random seating arrangement is allowed while listening to the khutbah. The imam stands on the minbar, usually 1 to 4 m higher than the main floor, and this usually permits line-of-sight contact between talker and listeners (see References M.2, M.6). Typically, the minbar is along the facing wall (quiblah) and includes a

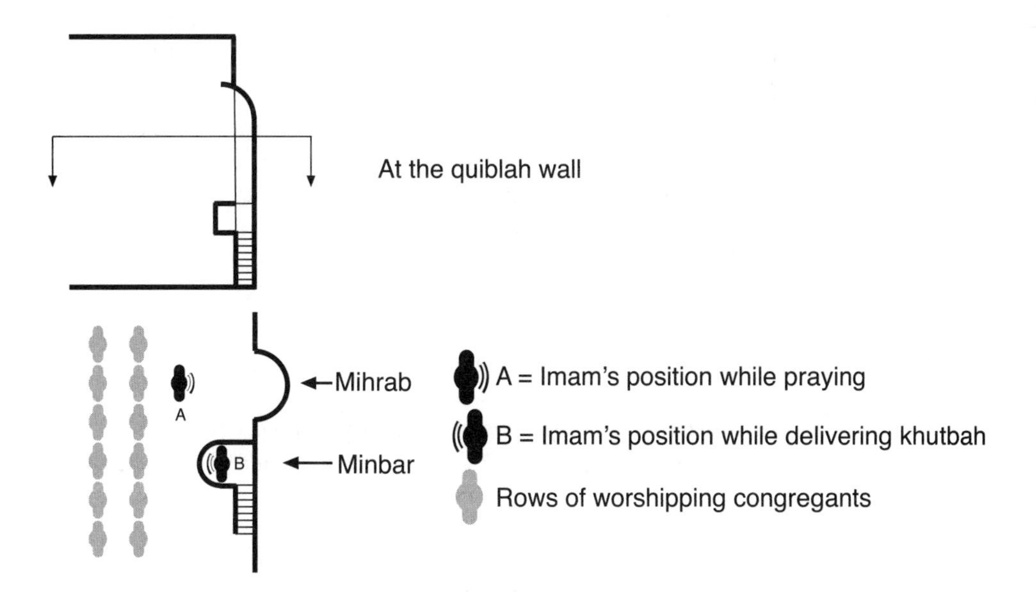

Figure M.3 The location of the imam at the quiblah wall while praying (a) or preaching (b).

little ceiling. The worshippers usually are on either a fully carpeted floor or small, personal rugs are distributed for their convenience. This sound-absorptive soft floor, along with the sound absorption/scattering due to the worshippers, means that the sound field will be less diffuse; the situation is acoustically somewhat similar to that in a fully seated concert hall. Note that group prayer may also take place in the outside courtyard and, in Iran, in the Shabestan (a cool, usually underground space).

M.3 MUSIC IN ISLAMIC WORSHIP

Instrumental music is not mentioned in the Koran, contrasting with the frequent mention of music and musical instruments in the Hebrew Bible or the Christian Old Testament (see Reference M.4). Unlike King David, Muhammad did not play a musical instrument. So, except for the voice (singing and chanting to emphasize the text's meaning), the Dervish tradition, and some Sufi mosques, music generally is not employed and musical instruments are not used in mosques. The chants and singing of the prayers and sacred texts can be musical, however.

The secular music of the Islamic world is highly sophisticated and worth studying and hearing, and possibly even learning to play, for western-trained ears and musicians. There is a similarity of some of the really elegant Muslim chanting, fine songs with variations and stanzas, to much of the music heard in the Sephardic synagogues.

In the Islamic world, some 25 instruments have been developed, independent of the instruments of western culture. These are fairly evenly divided between wind, string, and percussion. The number can be modified if one wants to account for minor changes in different traditions or changes of names. Some instruments were used throughout the Islamic world whereas others were restricted to one or two traditions. The history of Islamic music, the era of *The Great Musical Tradition* of the period of maximum expansion of the Arab empire, and the subsequent growth of the seven separate traditions—Persian, Andalasusian-Magrev, Turkish, Central Asian, Northern Indian and Pakistani, Bedouin, and Dervish—and today's situation are well worth study (see Reference M.4). The book, *The Mood of the 'ud* opens with Ibn Khaldun's statement:

> *"Song is an art that develops out of abundance—in a culture that has achieved a state of welfare, thus reaching the peak of indulgence" (see Reference M.4).*

However, speech and its intelligibility are the acoustical factors in mosque design, paralleling those in Orthodox Jewish synagogues and Christian churches (see References M.2, M.3).

M.4 MOSQUE ACOUSTICS AND SOUND SYSTEMS

M.4.1 Traditional Mosques

Domes are a tradition in Islam as are minarets. Many Muslims consider domes and minarets to be a necessary visual element of a mosque even though they may not be a part of the local culture. Most traditional mosques have dome ceilings inside. Some modern mosques, to keep to traditional mosque symbolism, have domes that are not represented inside the building by a corresponding domed ceiling.

Mosques traditionally had all hard interiors, except for the frequent use of prayer mats and rugs. Otherwise, at various times, alternative hard materials were used, including ceramic tile, glass, wood, and stone.

Figure M.3 shows the quiblah that is the wall on the side of the mosque facing Makkah, Saudi Arabia. It and its opposite wall are usually the longer pair of the four walls, but not always. Near or at the quiblah's center is the praying niche (mihrab), a recess in the wall, and to its right is the minbar that the imam stands on when he delivers the Friday and Holy Day khutbah. The mahfili is a special raised platform opposite the minbar from where the muezzin chants in response to the imam's prayers. These basic elements have not had changes since the prototype mosques of 622 and 623 CE even with the different styles of architecture discussed earlier (see References M.1, M.2, M.3). Figure M.4 shows a photo of the quiblah wall in the Jameh mosque, Isfahan, Iran.

The capacity of a mosque is determined by its floor area, at approximately 1 m^2 per person (see Reference M.2). When full, the floor space per worshipper may be considerably less.

Figure M.4 The quiblah in the Jameh mosque, Isfahan, Iran, with its mahfili (left), mihrab, and minbar (right) (photo: N. Amiryarahmadi).

Along with an emphasis on speech intelligibility, Islam also shares with Orthodox Judaism the separation of sexes in worship (see Reference M.2). However, this is accomplished more frequently by separate worship spaces for women rather than balconies or annexes as in Orthodox synagogues. These separate women's rooms are usually smaller than the mosque used by the men and may lack the quiblah and minbar (see Reference M.6).

Figure M.5 shows the interior of the Sokullu mosque, Istanbul, Turkey, that clearly shows the large volume associated with a large and high dome. Many historical mosques were designed along these principles and architects seeking to satisfy worshippers looking for tradition may be tempted to use similar large-volume approaches. These will, of course, have the same acoustical problems as the cathedrals in Christian worship. To overcome the large sound attenuation of the imam's voice over the rows of worshippers on the floor, the imam delivers his khutbah from the top of the minbar stairs.

Most of the sound from the imam at the top of the minbar will be feeding energy into the reverberant sound field. The worshippers will be listening more to reverberation than to direct sound and that leads to poor speech intelligibility. The minbar seldom has the local ceiling top associated with the pulpit as in Christian cathedrals so the acoustical situation may be worse, particularly in view of the fact that mosques have a quadratic floor plan (with the minbar close to the center of the quiblah) rather than the rectangular floor plan common for cathedrals.

To improve speech intelligibility, the reverberant sound must be reduced relative to the useful direct sound and possible early reflections. This can be done by using an appropriate reflector behind the imam or by using sound reinforcement systems. It is imperative that such systems have highly directional loudspeakers, arrays, or horns so that their sound mainly reaches the worshippers. Because of the sound absorption of the worshipper's clothing and the carpets, sound that reaches the floor area will be absorbed and will not reflect or scatter back to generate reverberation. This is an advantage compared to the situation in cathedrals where the floor area is acoustically reflective.

Except for the lack of any possibilities for the use of pew-back loudspeaker systems, the same sound system design considerations apply to mosques as those that apply to churches and synagogues. There are no special design considerations as in Orthodox synagogues. Use of electricity in its various forms is permitted on all days in Islam.

Chapter 10 shows several examples of the use of highly directional, electronically steerable loudspeaker arrays that can be mounted flush against walls or pillars. Because of the electronically generated sound-radiation pattern, the sound can be directed toward the worshippers in spite of the vertical mounting of the loudspeaker array. Figure M.6 shows the installation of such an array in the Grand Mosque in Abu Dhabi.

(a)

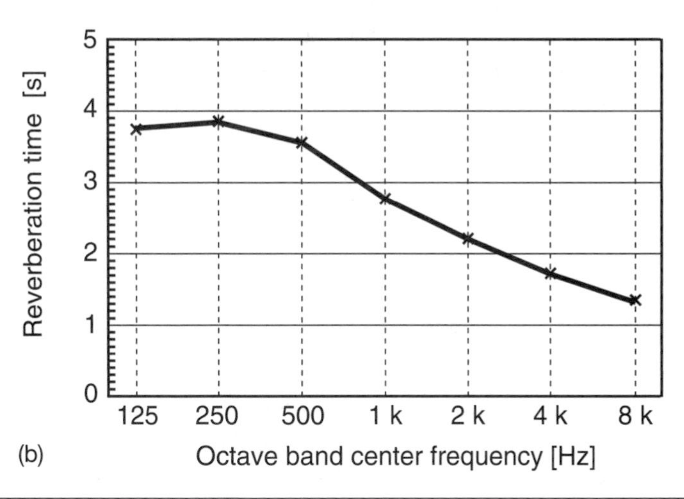

(b)

Figure M.5 (a) The quiblah in the Sokullu mosque, Istanbul, Turkey, with its mihrab and minbar. Master Sinan, architect (1572 CE) (photo: J.-H. Rindel). (b) The reverberation time in the Sokullu mosque, Istanbul, Turkey. Measured empty the volume is 5,700 m^3 (data: J.-H. Rindel).

Figure M.6 An electronic steerable Intellivox array installed on a pillar in the Grand Mosque in Abu Dhabi (photo: Duran Audio, www.duran-audio.co.uk).

During the installation of any loudspeaker system in a mosque, it is important to take into account the time delays between the direct sound and the amplified sound from the loudspeakers. Not only is intelligibility rather than audibility desired, but also that the imam's voice appears to come from the imam rather than the closest loudspeaker. To achieve this, the loudspeakers need to be fed suitably delayed signals so that the direct sound of the imam's voice reaches the worshippers before any amplified sound. Typically, the delays are set according to the distances between the minbar and each individual loudspeaker. For reasons of cost, large directional arrays will be mounted more sparsely than smaller commercial, mechanically tilted arrays.

Howl (acoustically generated feedback) is always a possibility in any sound amplification system. It is reduced by the use of three approaches: (1) directional loudspeaker arrays and horns, (2) close and directional microphones, and (3) electronic signal processing. These are all described in detail in Chapter 10. It is likely that some form of electronic feedback control will be necessary in the sound system.

Speech intelligibility will be hampered not only by reverberation but also by echoes due to the large size of traditional mosques. Echoes will be a problem for the imam also because the presence of echoes makes it more difficult to speak. Echoes will need to be reduced by some of the methods described in Chapter 7.

M.4.2 Contemporary Mosques

Most mosques' construction in the United States and Europe started after 1960 when Muslim immigrants began coming from the Middle East and many other places in the world. A large number of American and European Muslims are first generation immigrants (in the United States about one-third are African Americans).

Three styles have characterized American mosques (see Reference M.7): (1) imported styles in which mosques are constructed according to the tradition of the country of origin. These are conservatively styled mosques, featuring domes and minarets, for example, the mosque in the Islamic Center in Washington D.C. (Mario Rossi, architect, 1950); (2) adapted mosques that were built with the intention to mold traditional and new country elements. The mosque in the Islamic Cultural Center in New York (Skidmore, Owings, & Merrill, architects, 1991) and shown in Figure M.7; and (3) innovative mosques that were built in a new architectural style such as the mosque of the Islamic Society of North America, Plainfield, Indiana (Gulzar Haider, architect, 1981).

Mosques may have distinctly varying floor plans as shown in Figure M.8. It shows common floor plans used for smaller mosques (typically having a volume of less than 3000 m^3) (see References M.2, M.3).

However, innovative mosques typically retain the traditional rectangular floor space. Most new mosques will keep the traditional quiblah, minbar, and mihrab arrangement that one finds in traditional mosques.

Figure M.7 The interior of the Islamic Cultural Center in New York (photo: Imran Ali, www.imranali.name).

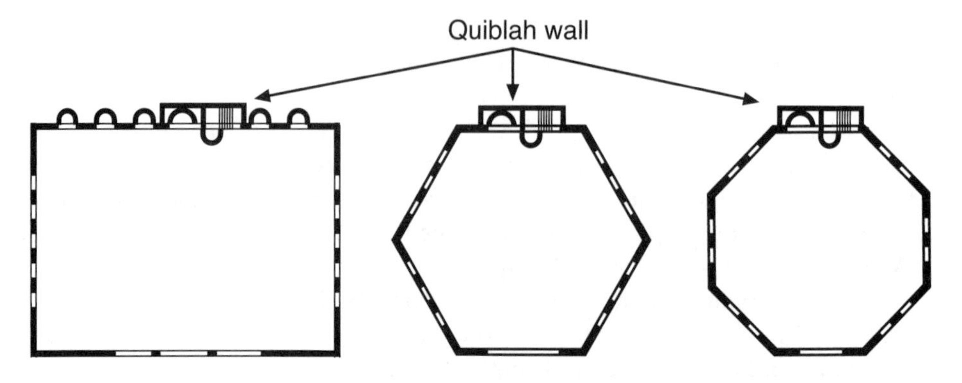

Figure M.8 Rectangular, hexagonal, and octagonal floor plans are common in smaller mosques.

In Europe as well, many new mosques are being built. A recent addition, the Ahmadiya mosque in Penzberg, Germany, shown in Figure M.9, has its interior split into two planes (men and women, respectively) joined at the quiblah by a two-story room.

Because women in the United States will participate in Islamic worship more than in traditional mosques outside of the United States, more area is set aside for them than in Saudi Arabia, for example. Figure M.10 shows the interior of the

Figure M.9 The interior of the Ahmadiya Mosque in Penzberg, Germany, showing stylized minbar and mihrab. Women's worship space is on the upper level and the men's are on the lower level (photo: Morten M. Lund).

Figure M.10 Interior of the DH03 mosque, Al-Doha, Saudi Arabia. Volume = 5105 m^3, interior ceiling height = 6.05 m. The sound system uses 12 column loudspeakers mounted at a height of 2.7 m (photo: Adel A. Abdou).

DH03 mosque, Al-Doha, Saudi Arabia. In this mosque, the women's praying area is a small part of the main praying area. It is also common in the Middle East, Africa, and Asia to have separate mosques for men and women.

Many newly built community mosques tend to be smaller than the traditional mosques discussed earlier and, thus, have a relatively low ceiling height. Because of this, the minbar typically will be nearly at floor level and that results in a poor transmission of the direct voice of the imam. Often the ceilings will feature beams or ventilation ducts that prevent them from effectively carrying early overhead sound reflections. These conditions, along with the noise generated by the ventilation system, make sound amplification systems a necessity. Such systems can be installed using the pillars of the mosque for support, due in part to the convenience of installation. However, a straight-forward ceiling-distributed system planned in conformance with Chapter 10 (see Figure 10.1c) would provide better results in nearly all cases with or without the use of signal delay for realism. Small loudspeakers on columns result in numerous listeners hearing sound at varying arrivals times from many loudspeakers, a problem minimized with ceiling loudspeaker directing sound vertically to the sound-absorbing congregation and prayer rugs. A small, low-ceilinged mosque should not require a sound system if room acoustics and noise control are handled properly, even though there are pillars present.

Extensive measurements of the acoustics of Saudi Arabian mosques have been done by Adel Abdou (see Reference M.2). He measured mosques of various sizes that had different floor plans. What he found in regard to the range of

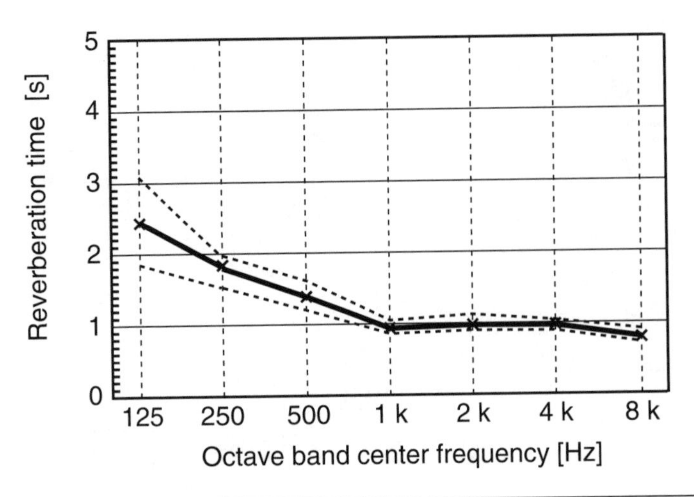

Figure M.11 Maximum, average, and minimum reverberation times $RT_{-5,\,-35}$ found in an investigation of 20 Saudi Arabian mosques (see Reference M.2).

reverberation times for empty mosques is shown in Figure M.11. Because of the relatively low ceiling heights and carpets on the floor these mosques generally have short reverberation times at mid and high frequencies. The *Journal of the Acoustical Society of America* and the *Journal of the Audio Engineering Society* have both reported on measurements of mosque acoustics (see References M.2, M.9, M.11).

As in many church and synagogue situations, the authors believe that steered line-source or column loudspeakers using digital signal processing, discussed in Chapter 10, can be used to improve intelligibility in many mosques that have large ceiling heights. Such loudspeakers can provide a measure of architectural freedom to the designer while assuring good intelligibility in new ones such as the installation shown in Figure M.6. For those large mosques that have low ceiling heights, systems using many small ceiling loudspeakers directed vertically towards the sound-absorbing congregation and carpets are recommended.

M.5 NOISE CONTROL CONSIDERATIONS

Noise control criteria for mosques are the same as for Christian and Jewish worship spaces, but frequently these criteria are not met in any of these worship spaces. Noise control remains a problem in many mosques despite the technology available to solve this problem (see Reference M.1). Noise generally is due to fans and other air conditioning and ventilation systems. Their design should be left to a competent contractor.

Figure M.12 The Nablus Road Mosque, Jerusalem, Israel (left). Part of the old mosque is at the right of the minaret, and the new mosque that is under construction is to the left. Note the reentrant-horn loudspeaker on the minaret balcony (photo: David L. Klepper). The minaret and cupola of the Sofia Mosque, Bulgaria (middle) (photo: Bruce Russell). The symbolic minaret of the Ahmadiya Mosque in Penzberg, Germany (right) (photo: Morten M. Lund).

M.6 MINARETS

In Muslim countries and in Israel, mosques have towers called *minarets* (Arabic for slender tower with balconies) analogous to the bell towers of many Christian churches and cathedrals. Their basic purpose is similar; to call people to worship. The minarets were also used for ventilation purposes, achieving a chimney effect. In principle, the minaret should be high enough for the voice to project well over the neighborhood. In contrast to Christian churches, the calling signal is always a human voice rather than bells or other musical instruments or sounds. Today, a uniform practice is the use of loudspeakers in these towers.

Loudspeaker and amplification equipment for these minarets can be quite primitive and of a low quality such as paging reentrant horns (as shown on the minaret to the left in Figure M.12) and packaged mixer-amplifiers. They can also be well-designed, high-quality, and high-fidelity outdoor systems. The use of one live voice and a microphone pickup appears universal. Chapter 10 explains how loudspeaker directional characteristics can be used creatively to compensate somewhat for the inverse square loss described in Chapter 2.

As mentioned, many Muslims consider the minaret and dome to be the key symbols of a mosque. In Western countries, minarets using loudspeakers likely would violate noise ordinances. Most modern mosques in Western countries will have minarets that are purely symbolic. An example is shown on the right in Figure M.12.

APPENDIX 1

The approximate reverberation time of a room can be calculated using Sabine's equation that gives us the approximate relationship between reverberation time, room volume, and sound absorption in the room. The equation, however, strictly applies only under specific conditions. The sound field of the room must be diffuse and have little sound absorption. Rooms that are small and that have extreme geometries (such as a long corridor or office landscape), or one in which the sound absorption is not evenly located (such as on one main surface) will lack diffusion. It is then necessary to apply physical or mathematical models and study the sound field in a more exacting way. For large rooms this can be done using ray-tracing or mirror image methods to calculate the distribution and decay of sound in the room. The difference in calculated reverberation time can be significant.

In many cases, however, involving ordinary rooms in which there is a reasonably even distribution of sound absorption, a distribution of sound-scattering objects and surfaces, or relatively little sound absorption, Sabine's equation will give useful results. Frequently, they will not differ by more than 10% from the value obtained by more advanced methods.

A simple example of the use of Sabine's equation for rooms:

$$RT \approx \frac{0.161V}{\sum_i A_i} = \frac{0.161V}{\sum_{i=1}^{N} \alpha_i S_i + \sum_{j=1}^{M} A_j + 4mV} \quad \text{(Metric units)}$$

We will use the metric version where $\Sigma \alpha S$ is the total sound absorption in metric sabins. One metric sabin is a square meter of total sound absorption, $\alpha S = 1.0$. In the equation, V is room volume in cubic meters, α_i is the sound-absorption coefficient of the various surfaces in the room (both the room and the objects in the room), and S_i is the area of those surfaces in square meters.

The sound absorption due to air is 4 mV metric sabins. The attenuation coefficient m can be obtained from the graph in Figure 3.4 by dividing the dB values in the figure by 434. As illustrated by that figure, the attenuation for frequencies below 2kHz is negligible. We assume that the room temperature is 20°C (68°) and the relative humidity is 50%. We further assume that all the other sound absorption can be represented by surfaces that have a certain sound-absorption coefficient.

We will calculate the reverberation time for the frequencies 125 Hz, 500 Hz, and 2 kHz. Before using Sabine's equation we should check that the absorption is sufficiently low by calculating the average sound-absorption coefficient that should be lower than 0.3 for Sabine's equation to be applied. We should also check that the frequency limit for the use of Sabine's formula, given by the Schroeder frequency, is well below than the lowest frequency of the lowest octave band. The Schroeder frequency is given by

$$f_S \approx 2000\sqrt{\frac{RT}{V}}$$

The lowest frequency of the 125 Hz octave band it is 125 Hz divided by the square root of 2, that is 125/1.4 ≈ 88 Hz.

A small basilica plan community worship space is a rectangular shoebox having an interior width of 20 m, a length of 30 m, and a height of 7 m. There are 40 m^2 of windows and niches on the two side walls, and the remainder of each wall has an area of 170 m^2. A pipe organ exposing 25 m^2 of pipework is installed in a gallery on top of the narthex. A 2 m wide carpet covers 20 m of the center of the nave. The front and side walls consist of concrete blocks, grout-filled and painted. The rear wall is covered by perforated wood panels on concrete. The ceiling is plaster on lath. The floor is wood on concrete. The seating consists of wooden benches that have seat cushions. The seating area is 320 m^2. Note that the absorption area needs to be increased by the sound absorption of its perimeter as discussed in Chapter 4. If the seating is assumed to be in two blocks, 8 m by 20 m on either side of the carpet in the nave, the effective seating area on each side will be 9 m by 21 m. The rear wall has a door of 5 m^2 opening up to the narthex. The room volume is 4200 m^3.

The Schroeder frequency f_S is approximately 30 Hz for this room. The table on the next page shows that the mean sound-absorption coefficients are less than 0.3 in all three octave bands. For the organ we have assumed a default sound-absorption coefficient of 0.5. The distribution of sound absorption in the room is uneven, however, most of the absorption is due to the seating area on the floor, so the applicability of Sabine's equation is in doubt, since the equation is derived assuming that the sound absorption is evenly distributed over the room surfaces. For a space of this size, one should also include the sound absorption by the air

Surface number	Element	Surface area (m^2)	Frequency					
			125 Hz		500 Hz		2 kHz	
			Alpha	Absorption area (metric sabins)	Alpha	Absorption area (metric sabins)	Alpha	Absorption area (metric sabins)
1	Front wall, painted concrete	140	0.01	1.40	0.02	2.80	0.02	2.80
2	Side wall, painted concrete, left	170	0.01	1.70	0.02	3.40	0.02	3.40
3	Side wall, painted concrete, right	170	0.01	1.70	0.02	3.40	0.02	3.40
4	Windows, left side wall	40	0.36	14.40	0.18	7.20	0.07	2.80
5	Windows, left side wall	40	0.36	14.40	0.18	7.20	0.07	2.80
6	Wooden benches occupied	378	0.57	215.5	0.75	283.5	0.90	340.2
7	Carpet in nave	40	0.08	3.2	0.39	15.6	0.48	19.2
8	Wood floor exposed	240	0.04	9.6	0.06	14.4	0.06	14.4
9	Rear wall, perforated wood panel	110	0.37	40.7	0.63	69.3	0.96	105.6
10	Door in rear wall	5	0.50	2.5	0.50	2.5	0.50	2.5
11	Organ on gallery	25	0.50	12.5	0.50	12.5	0.50	12.5
12	Ceiling	600	0.14	84.00	0.06	36.00	0.04	24.00
13	Absorption due to air humidity, 20°C, 50%RH, V=4200 m^3			0.00		0.00		32.00
	Total sound-absorption area			401.6		457.8		565.6
	Total room surface area	1958						
	Mean sound-absorption coefficient		0.21		0.23		0.29	
	Reverberation time (RT)			1.68		1.48		1.20

which affects the highest octave band values. (Values for the attenuation coefficient m in Sabine's equation may be found in Figure 3.4 (right hand scale) or, for a wider range, on page 4.5 in Reference 8.4.)

The table on page 273 shows that the reverberation time values vary substantially over frequency. They are reasonably good for music (although traditional organ music sounds better with much longer reverberation times), give good warmth, and the room in its assumed state would probably need a good directional sound-reinforcement system to enhance clarity and intelligibility of speech.

PART I REFERENCES AND NOTES

Chapter 1

To learn more about the fundamentals of sound and hearing—but at an engineering mathematical level—the following book is recommended:

1.1 Raichel, D. R., *The Science and Applications of Acoustics*, 2nd ed. New York: Springer (2006). ISBN: 978-0387260624.

To learn more at a beginner's level, we recommend:

1.2 Benade, A. H., *Fundamentals of Musical Acoustics*. Dover Publications. (1991). ISBN: 978-0486264844.

Another good reference is:

1.3 Pohlmann, K. and Everest, F., *Master Handbook of Acoustics,* 5th ed. McGraw-Hill/TAB Electronics (2009). ISBN: 978-0071603324.

Chapter 2

2.1 Fletcher, H., *The ASA Edition of Speech and Hearing in Communication.* Acoustical Society of America (1995 reprint of 1961 edition). ISBN: 978-1563963933.

2.2 Spoor, A., Presbyacusis values in relation to noise-induced hearing loss. *Audiology* 6, 48 (1967).

Two general books on hearing:

2.3 Yost, W. A., *Fundamentals of Hearing: An Introduction*, 5th ed. Emerald Group Publishing (2006). ISBN: 978-0123704733.

2.4 Green, D. M., *Introduction to Hearing*. New York: John Wiley & Sons (1976). ISBN: 978-0470151884.

Chapter 3

Two excellent references on general room acoustics are:

3.1 Cremer, L., et al., *Principles and Applications of Room Acoustics (Volume 1)*. New York: Applied Science Publishers (1982). ISBN: 0-853341133.

3.2 Kuttruff, H., *Room Acoustics*, 4th ed. London: Taylor & Francis (2000). ISBN: 978-0419245803.

Chapter 4

4.1 Bradley, J. S., "Predicting the Absorption of Pew Cushions." *Journal of the Acoustical Society of America*, vol. 92:4, p. 2470 (1992).

4.2 Beranek, L. L. & Hidaka, T., "Sound Absorption in Concert Halls by Seats, Occupied and Unoccupied, and by the Hall's Interior Surfaces." *Journal of the Acoustical Society of America*, vol. 104, pp. 3169–3177 (1998).

4.3 Hidaka, T., et al., "Relation of Acoustical Parameters with and without Audiences in Concert Halls and a Simple Method for Simulating the Occupied State." *Journal of the Acoustical Society of America*, vol. 109:3, pp. 1028–1042 (2001).

Data for many acoustical materials and constructions may be found in:

4.4 Doelle, L. L., *Environmental Acoustics*. Montreal: McGraw-Hill (1972). ISBN: 978-0070173422.

4.5 Fasold, W., et al., *Bau- und Raumakustik* (in German). Berlin: VEB Verlag für Bauwesen, (1987). ISBN: 3345001403.

4.6 Long, M., *Architectural Acoustics*. New York: Academic Press (2006). ISBN: 978-0124555518

Chapter 5

5.1 Beranek, L. L., *Music, Acoustics, and Architecture*. New York: John Wiley & Sons (1962). ISBN: 978-0471068679. (Updated with new material available as Beranek, L. L., *Concert Halls and Opera Houses: Music, Acoustics, and Architecture*, 2nd ed. Springer (2003). ISBN: 978-0387955247.

5.2 Houtgast, T., et al., "Predicting Speech Intelligibility in Rooms from the Modulation Transfer Function. I. General Room Acoustics." *Acustica*, vol. 46, pp. 60–72 (1980).

5.3 Mapp, P., "Measuring Intelligibility." *Sound and Communications*, (April 2002) pp. 56–68.

5.4 Mehta, M., Johnson, J. and Rocafort, J. *Architectural Acoustics: Principles and Design*. Prentice Hall (1998). ISBN: 978-0137937950.

Chapter 6

6.1 Savioja, L., "Modelling Techniques for Virtual Acoustics." PhD thesis, Helsinki University of Technology, Telecommunications Software and Multimedia Laboratory (1999).

6.2 Kleiner, M., Dalenbäck, B.-I., and Svensson, P., "Auralization-An Overview," *Journal of the Audio Engineering Society*, vol. 41:11, pp. 861–875 (1993).

Chapter 7

7.1 Egan, D. M., *Architectural Acoustics*. J. Ross Publishing (2007). ISBN: 978-1932159783.

7.2 Long, M., *Architectural Acoustics*. New York: Academic Press (2006). ISBN: 978-0124555518.

7.3 Davis, D. and Davis, C., *Sound System Engineering*. Focal Press (1987). ISBN: 978-0672218576.

7.4 Bolt, Beranek, and Newman were consultants for the Tanglewood stage shell and gave similar advice to Walter Holtkamp, Sr., for the organ installation at the Manhattan's Corpus Christi Roman Catholic Church.

7.5 Klepper, D. L., "Behind the Actor's Back." *Journal of the Audio Engineering Society*, vol. 14:3 (1966) also in *Sound Reinforcement, an Anthology*. New York: Audio Engineering Society (1978), p. E-19.

7.6 Klepper, D. L., "Tent shaped concert halls, existing and future." *Journal of the Acoustical Society of America*, vol. 124:1, pp. 15–18 (2008).

7.7 Klepper, D. L., "First Presbytarian Church, Stamford, Connecticut." *Journal of the Acoustical Society of America*, vol. 31:7, pp. 879–882 (1959).

7.8 Crist, E. V., III, "Acoustics in the Worship Space: Goals and Strategies." Dissertation at University of Cincinnati (1993). Available from University

Microfilms International, 300 N. Zeeb Rd., Ann Arbor, MI 48106, order #9329952.

Much information on the planning of room acoustics may be found in References [7.1] & [7.2]. An excellent reference on the history of rooms for music is:

7.9 Forsyth, M., *Buildings for Music*. Cambridge, MA: The MIT Press (1985). ISBN: 978-0262561877.

7.10 Good references on the organ and church acoustics are *Acoustics in Worship Spaces* ($2.00) and *Planning Space for Pipe Organs* (free), both available from the American Guild of Organists, 475 Riverside Drive, Suite 126, New York, NY 10015, USA.

For readers of German the following book will also be of interest: Meyer, J., *Kirchenakustik* (in German). Frankfurt: Verlag Erwin Bochinsky. ISBN: 3-923639414.

For helpful discussions on organs and organ room acoustics, we gratefully acknowledge Hans Davidsson (EROI, Eastman School of Music, Rochester, NY); Johan Norrback, Paul Peeters, and Joel Speerstra (all at GOart, Gothenburg University, Gothenburg, Sweden).

Chapter 8

Much of the information in this chapter is distilled from Reference [8.5].

8.1 Tocci, G. C., "Room Noise Criteria—The state of art in the year 2000." *Noise News International*, vol. 8:33 (2000).

8.2 Harris, C. M., "Acoustical Design of Benaroya Hall, Seattle." *Journal of the Acoustical Society of America*, vol. 110:66, pp. 2841–2844 (2001).

8.3 Beranek, L. L., "Concert Halls and Opera Houses: Music, Acoustics, and Architecture." Acoustical Society of America (1996). ISBN: 978-0387955247.

8.4 Harris, C. M., *Handbook of Acoustical Measurements & Noise Control (Third Edition)*. New York: McGraw-Hill (1998). ISBN: 1-56396-774-X

8.5 "Guide of the American Society of Heating and Refrigerating Engineers." *ASHRAE*. Atlanta, GA.

8.6 McQuay International (www.mcquay.com) HVAC Acoustic Fundamentals Application Guide AG 31-010.

Basic information on noise control can also be found in:

8.7 Beranek, L. L., ed., *Noise and Vibration Control*. Institute of Noise Control Engineering. Washington, DC (1988). ISBN: 0-962007209.

Chapter 9

Basic information on sound isolation may be found in:

9.1 Beranek, L. L., ed., *Noise and Vibration Control,* Institute of Noise Control Engineering (1988). ISBN: 978-0962207204.

9.2 Hopkins, C., *Sound Insulation.* Butterworth-Heinemann (2007). ISBN: 978-0750665261.

9.3 Harris, C. M., *Handbook of Acoustical Measurements & Noise Control (Third Edition),* New York: McGraw-Hill (1998). ISBN: 1-56396-774-X

9.4 Watters, B. G., "Transmission Loss of Some Masonry Walls." *Journal of the Acoustical Society of America*, vol. 31:7, pp. 898–911 (1959).

Chapter 10

10.1 Davis, D. & Davis, C., *Sound System Engineering,* 2nd ed., chapter 10. Indianapolis: SAMS (1987). ISBN: 978-0240803050.

10.2 Eargle, J. and Forman, C., *Audio Engineering for Sound Reinforcement,* chapter 2. JBL Pro Audio Publications (2002). ISBN: 978-0634043550.

10.3 Klepper, D. L., "Sound Systems for Reverberant Spaces for Worship." Journal of the Audio Engineering Society, (August 1970) also in *Sound Reinforcement, an Anthology.* New York: Audio Engineering Society (1998), pp. E-46–56.

10.4 Long, M., *Architectural Acoustics.* New York: Academic Press (2005), p. 715. ISBN: 978-0124555518.

10.5 Eargle and Forman, op. cit., chapter 21.

10.6 Eargle and Forman, op. cit., pp. 28–29.

10.7 Haas, H., "Über den einfluss eines einfachechos auf die Hörsamkeit von sprache." *Acustica*, vol. 1 (1951), pp. 49–58.

10.8 Egan, D. M., *Architectural Acoustics.* J. Ross Publishing (2007), p. 374. ISBN: 978-1932159783.

10.9 Eargle and Forman, op. it., p. 217.

10.10 Eargle and Forman, op. cit., p. 320.

10.11 Klepper, D. L., "A Different Angle." *Sound and Video Contractor* (January 1999).

10.12 Harvard Memorial Church, located within Harvard Yard. University congregation supplemented by Cambridge residents. Traditional mainstream interdenominational Protestant services, excellent student choir, strong music program, including regular concerts and organ recitals. Reverberation time (*RT*) varies with occupancy and can be as low as

1 sec with a full church, but this doesn't prevent excellent congregational singing and general service music. (Bolt, Beranek, and Newman were responsible for coating the ceiling to be sound-reflecting, ca. 1964.)

10.13 Congregation Beth El, constructed in 1973, opened with a successful concert by the Detroit Symphony Orchestra. Reverberation time is 2 sec and has close to ideal concert-hall acoustics. Reform congregation with more congregational participation than classic reform but, otherwise, close to traditional mainstream reform practice. Architect was Minora Yamasaki.

10.14 North Shore Congregation Israel, Glencoe, Illinois. Main building illustrated opened in 1964. Beautiful synagogue campus located on wooded shore of Lake Michigan. Casavant Frerres organ, designed by Lawrence Phelps, in rear of the balcony, also the usual location for excellent professional choir. Strong music program with recitals and concerts (www.nsci.org).

10.15 Princeton University Chapel. University congregation supplemented by Princeton residents. Traditional mainstream interdenominational Protestant services, excellent student choir, strong music program, including regular concerts and organ recital.

10.16 Davis and Davis, op. cit., pp. 338–341.

10.17 Church of the Good Shepherd, Erie, Pennsylvania. Klepper, Marshall, and King sound system. Episcopal Church.

10.18 Klepper, D. L., "The Acoustics of St. Thomas Church Fifth Avenue." *Journal of the Audio Engineering Society*, vol. 43:7/8 (July/August 1995), pp. 599–601, also in "Sound Reinforcement, Anthology II," *Journal of the Audio Engineering Society* 1995, pp. 292–294.

St. Thomas Church was constructed in 1913; Cram, Goodhue, and Ferguson, architects. Its facilities include a boarding choir school. Yearly performances of Handel's *Messiah* are an additional important contribution to the New York musical scene, and there are many other concerts as well as organ recitals. The chancel organ was originally built by Ernest Skinner, 1913, and then rebuilt by Aeolian Skinner, 1956, Gil Adams, 1968, and Mann and Trupiano, 1980–1989. The rear gallery organ is by Taylor and Boody, 1996. Reverberation time is approximately 3.5 sec with typical occupancy and rising bass characteristics as is appropriate for most music performed in the church and its cathedral-like appearance. Nearly all the artificial porous stone, sound-absorbing tile has been sealed with multiple coats of sealer to achieve the desired reverberation characteristics.

10.19 (10.3 ibid). The National Presbyterian Church was built in 1968, Harold Wagoner, architect. The organ is by Aeolian Skinner, and the church has an excellent music program with traditional services. Reverberation time is 2 sec occupied, again appropriate for most music performed in the church.

10.20 Tappan, P. and Ancha, R., "Invisible Reinforcement with 350 Microphones." *Journal of the Audio Engineering Society*, (June 1978), also in "Sound Reinforcement, an Anthology," *Journal of the Audio Engineering Society*, (1996), pp. E-40–45.

10.21 St. Louis Priory Chapel, St. Louis, Missouri. Now a boys' school chapel and a chapel for Benedictine monks as well as a parish church for local laymen. Radetsky organ, choir of boys, traditional Catholic services, and good music program. Bolt, Beranek, and Newman sound system design illustrated.

10.22 Klepper, D. L. and Steele, D., "Constant Directional Characteristics from a Line-Source Array." *Journal of the Audio Engineering Society* (July 1963), also in "Sound Reinforcement, an Anthology," *Journal of the Audio Engineering Society*, p. D-21–25. As noted, this technology has been superseded by steerable array loudspeakers.

10.23 Klepper, D. L., "The Distributed Column Sound System at Holy Cross Cathedral, Boston, the Reconciliation of Speech and Music," *Journal of the Acoustical Society of America*, vol. 99:1 (1996), pp. 417–425. The cathedral has one of the best preserved and restored Hook late-19th century organs. Mid-frequency reverberation time approximately 3 sec occupied, as appropriate.

10.24 Parkin, P. H. and Taylor, J. H., "Speech Reinforcement in St. Paul's Cathedral," *Wireless World* vol. 58:2 (February/March 1952), p. 58 and vol. 58:3, p. 109. Also in *Journal of the Audio Engineering Society*, vol. 54:1/2 January/February 2005), pp. 67–71. This latter paper presents the work that started the revolution in worship space acoustics to combine high intelligibility for speech with long reverberation times for liturgical music.

10.25 The steerable column loudspeakers now available take away much of the needed design work on paper, but the application of directional horn loudspeakers still requires this effort. John Prohs and David Harris PHD software for loudspeaker installations was developed at Ambassador College. This was the first reliable directional loudspeaker computer design program and is still appreciated for its simplicity. It replaced the light analogy technique originated by Ewart Weatheril and Wilfred Malmlund, (see Weatheril, E. and Malmlund, W. "An Optical Aid for

Designing Loudspeaker Clusters." *Journal of the Audio Engineering Society,* 12:1, pp. 57–61, 1965) and "Sound Reinforcement, an Anthology," *Journal of the Audio Engineering Society*). All values are entered manually, although reverberation time and average absorption coefficient values are interdependent and only one need be entered. The program does provide for calculating the average absorption coefficient from areas and their local coefficients. One also can choose to list an absorption coefficient for the areas within loudspeaker coverage and another for those receiving basically reverberant energy only, also an NC background noise level if considered a factor (above NC-35). The software is described in (See Reference 10.1). Some features have been added since.

An interesting loudspeaker-room interface situation exists at the Christian Theological Seminary (CTS) Chapel, Indianapolis, Indiana, shown in the upper figure on page 283.

The drawings are not to scale and are intended as schematic only. The upper sketch on the left shows system and seating as designed. The lower sketch on the left represents the *church-in-the-round* use that is frequent but was not anticipated in design. To reduce the high reverberant levels resulting from a bare marble floor area in the primary coverage pattern of the central loudspeaker system, a square of heavy carpet with an underlay is placed in the center of the *square.* A wide variety of educational, religious, music, and theater programs are available at CTS and many are open to the general public. The chapel also has a Holtkamp organ. The mid-frequency reverberation time is 2 sec with normal full occupancy.

10.26 Davis and Davis, op. cit., p. 467.

10.27 The lower figure on page 284 shows a circuit, developed by Peter Tappan, allowing a paging signal to bypass the effect of local room loudspeaker line volume controls.

Allowance should be made for the additional load of the resistive room volume controls on the amplifier output. If inductive controls are used, high-pass filtering of the input to the power amplifier is essential; 80Hz or 100Hz is recommended.

10.28 Griesinger, D., "Improving Room Acoustics through Time-Variant Synthetic Reverberation." *Journal of the Audio Engineering Society,* 90th Convention (1991).

10.29 Chapter 3 explains the terms *RASTI* and *ALCons.* Also refer to Snow, W. B., "Articulation Loss of Consonants as a Criterion for Speech Transmission in a Room," *Journal of the Audio Engineering Society* (January 1969), also "Sound Reinforcement, an Anthology." *Journal of the Audio Engineering Society,* pp. A-22–27 and Klein, W., "Articulation Loss of Consonants as a Basis for the Design and Judgment of Sound Reinforcement Systems."

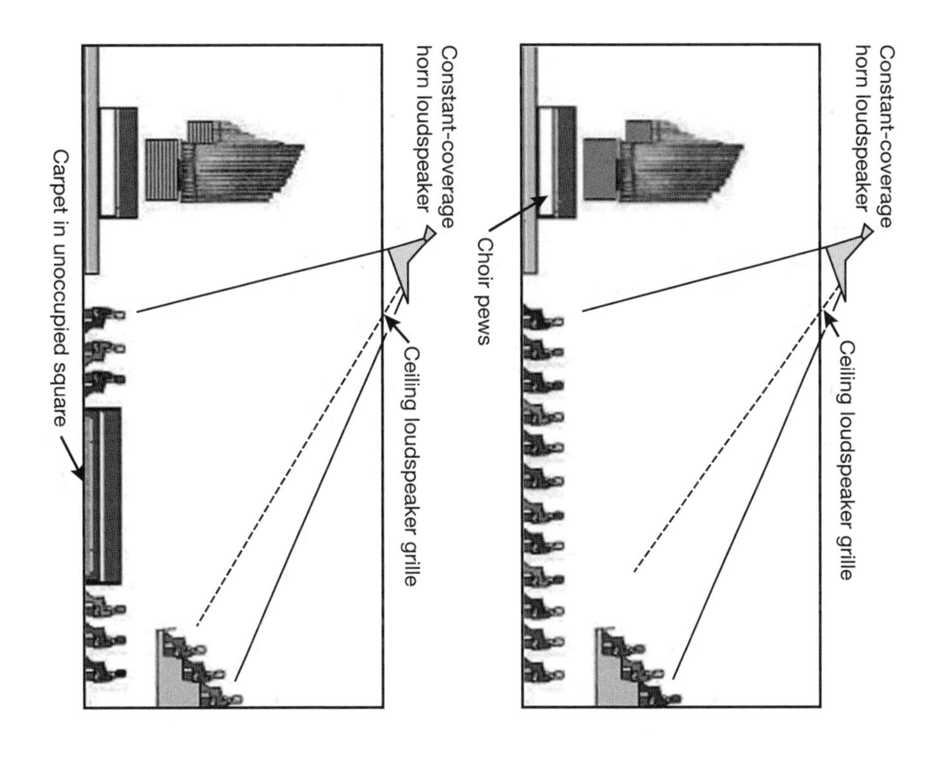
Constant-coverage horn loudspeaker
Choir pews
Ceiling loudspeaker grille
Constant-coverage horn loudspeaker
Ceiling loudspeaker grille
Carpet in unoccupied square

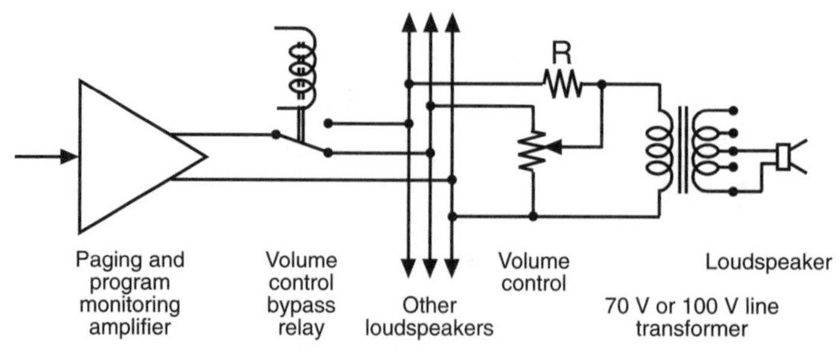

Journal of the Audio Engineering Society (December 1971), also an anthology, *Journal of the Audio Engineering Society,* pp. E-28–29. For the Patronis Method of adjusting levels in loudspeaker systems, Patronis, E. T. Jr. and Donders, C., "Central Cluster Design Technique for Large Auditoriums." *Journal of the Audio Engineering Society*, vol. 30:6 (June 1982), pp. 407–411, also "Sound Reinforcement." *Journal of the Audio Engineering Society,* vol. 2 (1996), pp. 121–125; specifically the paragraph following the equations, p. 410/124. The John Prohs–David Harris PHD computer program is described in (see Reference 10.1) pp. 535–537. Some features have been added.

10.30 Kaye, D. and Klepper, D. L., "Sound System Specifications." *Journal of the Audio Engineering Society* (April 1962), also in "Sound Reinforcement, an Anthology," pp. A-1–4, Reference [10.1], also an ongoing series of articles by Peter Mapp in the journal *Sound and Communications* (2007, 2008, 2009).

10.31 Klepper, D. L., "Use of the L. G. Marshal-Crown-Techron ELR Program for Adjusting Digital Delay Units in Sound Reinforcement Systems." *Journal of the Audio Engineering Society*, vol. 43:11 (1995), pp. 942–945.

10.32 Marshall, L. G., "An Acoustics Measurement Program for Evaluating Auditoriums Based on the Early/Late Sound Energy Ratio." *Journal of the Acoustical Society of America*, vol. 96 (1994), pp. 2251–2261.

10.33 Klepper, D. L., *Thirty Years of Sound System Progress.* Syn-Aud-Con Newsletter, Fall 1981.

PART II REFERENCES AND NOTES

Examples of many worship spaces may be found in:

Lubman, D. and Wetherill, E. A., "Acoustics of Worship Spaces." New York: American Institute of Physics, (1985). ISBN: 0-883184664

Synagogues

Readers desiring to visit a synagogue are urged to do so on Friday evenings or Saturday mornings. It is often advisable to contact the synagogue office a few days in advance. Most synagogues have a prayer book, Siddur, that has the Hebrew texts and English translations on facing pages so that, with the help of a friendly congregant, following the liturgy should be possible for those without Hebrew knowledge. Dress should be modest and semiformal, wear a head covering such as a skullcap, yarmulke, (or a Fedora in winter or a Panama in summer), stand when the congregation stands, and sit when they sit. Do not bring food into the synagogue at any time and do not write. Also, remember to switch off your personal phone before entering the synagogue. You will be welcomed and find the experience both interesting and enjoyable.

S.1 Tachau, W. G., "The Architecture of the Synagogue," The American Jewish Year Book (1926/1927).

S.2 Kohl, H. and Watzinger, C., *Antike Synagogen in Galilea.* Leipzig 1916 (reprinted by Kedem Publishing House, Jerusalem, 1973).

S.3 *Encyclopedia Judaica*, entry on "Synagogue."

S.4 *Encyclopedia Judaica*, entry on "Music." For a detailed description of Orthodox Ashkenazi or Hassidic worship practice in English, the reader is referred to the *The Complete Artscroll Siddur* (two versions for the two traditions), The Mesora Publications Limited, Brooklyn, NY, 1984. For Sephardic practice in English, the *de Sola Pool Siddur*, Union of Sephardic Congregations, New York, first printing, 1941.

S.6 The Touro Synagogue in Newport, Rhode Island, was constructed in 1763 for a local Sephardic Congregation whose numbers were increased by summer residents from Congregation Shearith Israel in New York and Mikvah Israel in Philadelphia. Today the building is owned by Shearith Israel and is leased to the local Ashkenasi Congregation, Jeshuat Israel. Visitors are welcome, and it was added to the list of U.S. National Historic Landmarks in 1947.

S.7 Congregation Young Israel of the Main Line, Bala Cynwyd, Pennsylvania (www2.yiml.org).

S.8 The old study hall/synagogue at Yeshivat Beit Orot. The view is typical of many Yeshiva synagogue/study halls.

S.9 Congregation Belz, Jerusalem, Israel. Building opened in 2000 after 15-year construction period. Congregation originally from Belz, Ukraine, moved to Israel in 1945. Suggest contact in advance to attend a Sabbath or holiday service.

S.10 Judiska församlingen i Göteborg (www.judiskaforsamlingen.se).

S.11 Central Synagogue, New York, www.centralsynagogue.org. Open for visitors on most days. Designed by architect Henry Ferbach and completed in 1872. Congregation, result of merger of two congregations, started in 1839 and 1846. Thoroughly renovated and rebuilt 1998 to2001. Hardy Holtzman Pfeiffer were the architects and Jaffe-Scarboro were acoustical consultants. Added to the list of U.S. National and Historic Landmarks in 1975.

S.12 Temple Israel, Albany, New York (www.templeisraelalbany.org).

S.13 Congregation Shaarey Zedek, Southfield, Michigan (www.shaareyzedek.org). Congregation founded in 1861. Construction of current building completed in 1962. Architects were Percival Goodman and Albert Kahn; Acoustics, Bolt, Beranek, and Newman. Reverberation time (RT) approximately 1.7 sec at mid-frequencies, full, and with side walls closed; reduced to 1.6 sec with side walls open and the social halls in use for synagogue overflow seating.

S.14. D. Klepper, "A Different Angle." *Sound & Video Contractor* (January 1999).

S.15 Congregation Shearith Israel is North America's oldest Jewish congregation. It was established by the twenty-three Sephardic Jews who arrived in New Amsterdam in 1654, fleeing from the newly imposed Inquisition in Brazil, specifically the town of Recife. Its present building is its fifth, built in 1897, in the Greek Revival style of architecture (Emeritus Rabbi Marc Angel prefers neoclassical), contains two worship spaces, the *Little Synagogue*, modeled after the original Mill Street Synagogue of 1730 and the subsequent Crosby Street of 1834, with furniture and fixtures from those two synagogues, and including a lamp, centuries older, from Holland, and the Main Synagogue. Both synagogue rooms are basically rectangular shoebox in shape (See L. L. Beranek, "Concert Halls and Opera Houses, How They Sound." *Acoustical Society of America* (1996), p. 28, *Architectural Acoustics*, Marshall Long, Elsevier Academic Press (2006), pp. 26–32; and Cyril M. Harris, "Acoustical Design of Benaroya Hall, Seattle." *Journal of the Acoustical Society of America*, vol. 110:6, December 2001) pp. 2841–2844, both have hard wall and ceiling finish materials, mostly plaster, but the Little Synagogue is close to square whereas the much larger Main Synagogue has a more typical rectangular plan.

There are reasons to visit this synagogue, particularly on a Friday evening for Sabbath evening (Erev Shabbat, Shabbat Ma'ariv) and Saturday morning (Shachrit Shabbat) services. The room itself is one of the most beautiful worship spaces in North America with Tiffany stained-glass windows. Most of the Sabbath Liturgy is sung in extremely beautiful melodies, mostly traditional, and with the participatory congregation that includes many with beautiful voices, led by a professional male choir, with an expert director, Leon Hyman, an excellent chazzan cantor (Hazzan), Rabbi Ira Rohde, and rabbis who themselves have good singing voices, including Rabbi Hayyim J. Angel and his father, Emeritus Rabbi Marc D. Angel. In contrast to most Sephardic congregations, in North America, Israel, and worldwide, Western composers such as Rossi and Sulzer are represented rather than exclusively the Eastern melodies akin to Islamic chant heard from the minarets of mosques. The experience can be compared with attending an expertly performed High Latin Mass. Decorum is absolute in contrast to the easy informality and great exuberance of other Sephardic congregations. There is even attention to choreography with those congregants receiving honors who are taught the necessary steps. Of course, all is in Hebrew except for the Saturday morning Drash that is in easily understood English and can be compared with the sermon or homily of a Christian denomination. This is typical of

North American Sephardic congregations, as is the high level of congregational participation and attention that generates a spiritual atmosphere in all these congregations.

The room shares all the characteristics but one of a typical shoebox-shaped classical concert hall (Beranek ibid.). The ceiling is hard plaster with coffering. The side walls have the kind of *regular-irregularities* that typify most concert halls. The U-shaped and tiered balcony along the side walls and rear wall is normally used for the seating of ladies whereas men occupy the side-tiered seating on the main floor. A center platform (bimah) is used for much of the liturgy, including the readings from the Five Books of Moses (Torah), the associated Prophetic Readings (Haftara), the Book of Esther (Megilla) on Purim, and other sacred readings, all facing the front or east wall. At present there still is no special provision to assure high speech intelligibility for those behind the back of the person chanting at this location, which means that speech intelligibility is slightly less for them than those located between this center location and the front wall. This is not a problem for the congregants, because all are expected to follow the service and the readings with their eyes in the books in the pew-back racks. They do expect the sounds of the chanting to be beautiful and accurate.

Even before upgrading an existing sound system to subtly reinforce speech of the Drash, intelligibility was sufficiently high because the Rabbi who is speaking faces the congregation, receives some low-frequency reinforcement from the hard front wall behind him, and further reinforcement for listeners toward the rear by reflected energy from the coffered ceiling and the sound-reflecting balcony faces. Apparently, for nearly 100 years, the additional intelligibility that could have been provided by a pulpit canopy over the Drash location has not been felt necessary.

The reverberation time of the space with typical occupancy is about 1.7 sec with some music having an audible *tail*. The choir is located in a front balcony, higher than the U-shaped balcony for women's seating, with a hard ceiling overhead that permits choir members to hear each other and that also aids projection to the congregation. This balcony also contains George Gershwin's reed organ, left there when he moved from New York to Hollywood. It has been used for weddings in the synagogue, but no musical instruments, other than the Shofar on appropriate holidays, is permitted on the Sabbath and Holy Days. It was playable when last visited in 1995.

Except for the absence of a pulpit canopy, the acoustics of this space were probably as close to optimum as possible in the years before the intelligibility revolution pioneered by Parkin's and Taylor's work in London's St. Paul's Cathedral in the mid-29th century ("Speech Reinforcement in St. Paul's Cathedral." *Wireless World*, vol. 58:2, p.58 and vol. 58:3, p. 109, P. H. Parkin and J. H. Taylor, (February and March 1952), reprinted in the *Journal of Audio Engineering Society*, vol. 54:1/2, (January/February 2006). Also see *Architectural Acoustics*, Marshall Long, Elsevier Academic Press (2006), p. 715, second paragraph. For the older view of balancing speech and music acoustics, see *Acoustical Designing in Architecture*, Vern O. Knudsen and Cyril Harris, John Wiley and Sons, New York, 1950.) Acceptable or satisfactory acoustics, with the implication that one should be a public speaker and project, along with moderate liveliness for music, was probably a far better solution than either musical reverberation and enough to truly produce *mystery* along with close to unintelligible speech, or conversely high speech intelligibility for all speakers but dry acoustics for all music. The architect, Arnold Brunner, was either lucky or experienced.

The de Sola Pool prayer book, Siddur, Book of Prayers, has the Hebrew texts and English translations on facing pages, so with the help of a friendly congregant, following the liturgy should be possible for those without Hebrew knowledge. Dress should be modest and semi-formal, wear a head covering that may be a Fedora in winter or a Panama in summer, not necessarily a yarmulke, stand when the congregation stands and sit when they sit. Do not bring food into the synagogue at any time, and do not write or use a personal phone on the Sabbath. You will be welcomed and find the experience both interesting and enjoyable.

Much of this information is from personal experience, but historical matters are drawn from the book *Remnant of Israel*, Rabbi Marc D. Angel, Riverside Book, New York, NY, (2004), available from the congregation located at 2 West 70th Street, offices at 8 West 70th Street, New York, NY, email www.shearithisrael.org. The famous Touro Synagogue in Newport, Rhode Island is North America's oldest existing synagogue building, and Shearith Israel is the oldest congregation.

When the Zomet Institute, an organization under the Chief Israeli Sephardic Rabbi, provided guidelines for Shabbat speech reinforcement, an existing system was modified to meet those guidelines, and equalization was added to provide what was felt to be an improvement in drash speech intelligibility from satisfactory to excellent and without any obvious notice that amplification was used. The public was made aware of the use of amplification in a short article by Rabbi Hayyim Angel in the Synagogue Bulletin, November-December 2002. This system has been further improved by Dan Clayton, acoustical consultant, and Norcon

Communications, contractor, and still employs four column loudspeakers, now of the state-of-the-art *computer-steered-array* type, inconspicuous on the front wall, two on each side of the central Ark, an upper one serving the balcony, and a lower one serving main floor seating, For a discussion of the Zomet approach, see *Cross-Roads,* vol. II, IV, and V, Zomet Institute, Gush Etzion, Israel.

For the non-Jew visiting Jerusalem and wishing an introduction to Jewish practice, a most economical and pleasant Friday evening can be in the Ma'alot Daphne neighborhood, easily reached by public transportation before sundown, at one of the many neighborhood synagogues. Dress may be casual or formal, but must be modest, and a head covering is essential in the synagogue. But do not wear a tallit (prayer shawl, mostly or completely white). Sit when the congregation sits and stand when they stand. At the time of writing, afterward, you could have dinner at Rabbi Mordecai Machlis's home, which will be only a few minutes walk away from the synagogue. Open house for people of all faiths and it is a neighborhood landmark. Often there are over 100 guests. (Ma'alot Daphne, Building 137). Taxis or walking are suggested for the return.

Acknowledgments

The authors thank Dr. David Lubman, Rabbi Dr. Daniel Katz and Rabbi Michoel Danow for comments and advice. The authors are also grateful to acoustics consultant David Epstein of Efrat, Israel, for arranging possibilities for us to measure the acoustics of some Israeli synagogues.

Rabbi Marc D. Angel, Rabbi Emeritus of North America's oldest Congregation, Shearith Israel of New York, and former President of the Rabbinical Council of America. Read the manuscript of Reference 13, which furnished one basis for this book.

Churches

C.1 Metzger, M. and Beaumont, M., *History of theLiturgy: the Major Stages.* Collegeville, MN: Liturgical Press, (1997), p. 31. ISBN: 978-0814624333.

C.2 Ibid. p. 37.

C.3 In *The Ecclesiastical History*; Ibid. p. 37.

C.4 Ibid. p. 38.

C.5 Ibid. p. 40; *Catechism of the Catholic Church* #1345 ff.

C.6 Ibid. p. 31.

C.7 Ibid. pp. 120–122.

C.8 Association of Religion Data Archives, http://www.thearda.com/maps Reports/reports/mainline.asp

C.9 Segler, F. M. and Randall Bradley, C., *Christian Worship: Its Theology and Practice*, 3rd ed. B&H Academic (2006), pp. 36–38. ISBN: 978-0805440676 Also Wainwright, G. and Westerfield Tucker, K. B., *The Oxford History of Christian Worship*. Oxford University Press (2006), pp. 548–550. ISBN: 0195138864.

C.10 We are contrasting spaces with reverberant or non-reverberant acoustics, because among the many acoustical attributes that describe a listening space (e.g., reverberation, acoustical intimacy, sound strength, spectral balance, and spaciousness), the dominant and fundamental parameter after loudness is reverberation time.

C.11 Gill, G. D., *Music in Catholic Liturgy: A Pastoral and Theological Companion to Sing to the Lord*. Hillenbrand Books Chicago/Mundelein (2009), p. 26.

C.12 Torres, R. R. and Galiè, K., "Music and Acoustics for Catholic Churches History and Modernity." Presentation at the meeting of the Acoustical Society of America. Salt Lake City, Utah, June 8, 2007.

C.13 USCCB Music Subcommittee of the Committee of Divine Worship. *Sing to the Lord: Music in Divine Worship*. Washington, DC: United States Conference of Catholic Bishops (2007), p. 88. ISBN: 978-1601370228.

C.14 International Committee on English in the Liturgy, *General Instruction of the Roman Missal*, 3rd ed. Washington, DC: USCCB Publishing (2002), p. 393. ISBN: 978-1575555430.

C.15 USCCB Music Subcommittee of the Committee of Divine Worship. *Sing to the Lord: Music in Divine Worship*. Washington, DC: United States Conference of Catholic Bishops (2007), p. 101. ISBN: 978-1601370228.

C.16 Ibid., p. 102.

C.17 Ibid., p. 103.

C.18 Ibid., p. 104.

C.19 Fleisher, D., Pastoral Music. *Journal of the National Association of Pastoral Musicians* (April/May 1997), pp. 1–8.

C.20 USCCB Music Subcommittee of the Committee of Divine Worship. *Sing to the Lord: Music in Divine Worship*. Washington, DC: United States Conference of Catholic Bishops (2007), p. 32. ISBN: 978-1601370228.

C.21 All forms of worship today are contemporary by definition, but for lack of a universally agreed-upon name, the contemporary worship style refers specifically to the described characteristics in this section.

C.22 Cirillo, E. and Martellotta, F., *Worship, Acoustics, and Architecture*. Essex, UK: Multi-Science Publishing (2006). ISBN: 978-0906522448.

C.23 Martellotta, F., "Identifying Acoustical Coupling by Measurements and Prediction-Models for St. Peters Basilica in Rome," Journal of the *Acoustical Society of America*, vol. 126:3 (2009).

Other references are:

Association of Religion Data Archives [online]. Available from the World Wide Web: <http://www.thearda.com/mapsReports/reports/mainline.asp>

Catechism of the Catholic Church (in English). Libreria Editrice Vaticana (1997). ISBN: 978-0892435661.

Gros, J., et al., *Introduction to Ecumenism*. Mahwah, NJ: Paulist Press International (1998). ISBN: 978-0809137947.

Hoppough, G. J., *Liturgical Studies* (course notes), Blessed John XXIII National Seminary, Newton, MA (2009).

Hoppough, G. J., *Theology of Eurcharist* (course notes), Blessed John XXIII National Seminary, Newton, MA (2009).

Lee, R. W., *Acoustical Analysis and Multiple Source Auralizations of Charismatic Worship Spaces.* M.S. Thesis, Rensselaer Polytechnic Institute, Program in Architectural Acoustics (2003), Troy, NY.

Martellotta, F., "Subjective study of preferred listening conditions in Italian Catholic churches," *J. Sound and Vibration* 317 (2008), pp. 378–399.

Pope Paul VI. *Constitution on the Sacred Liturgy* (Sacrosanctum Concilium). United States Catholic Conference (1963). ISBN: 978-1555863036.

Pope Benedict XVI. *The Sacrament of Charity: On the Eucharist as the Source and Summit of the Church's Life and Mission.* Pauline Books & Media (2007). ISBN: 978-0819871039.

Rice, H. L. and Huffstutler, J. C., *Reformed Worship*. Louisville, KY: Geneva Press, p. 100 (2001). ISBN: 978-0664501471.

A Note on Acoustics in Mormon Churches

The worship in the Church of the Latter Day Saints (Mormon Church) is not different from mainstream North American Protestant churches. It has hymns, recital of texts, and sermons or homilies, and some of the Mormon worship spaces are not markedly different from mainstream Protestant churches either, but there are some differences that should be noted. Many of the buildings are constructed to standard plans that are variants of the plan and sections shown in Figures 9.18 and 10.11. They basically consist of a rectangular chapel primarily for worship and a somewhat larger space called the cultural hall. The two are separated by sound-rated movable partitions with the sound-lock chamber between them contributing to enough sound isolation to allow simultaneous activities in both spaces without acoustical interference. Ceilings in both spaces are hard and sound-reflecting, permitting both spaces to be used for music events, separately or joined. Sound absorption in the chapel is provided primarily by its occupants, whereas the rise in reverberation time when empty is controlled by upholstered seats, usually permanently installed. Design reverberation times are largely controlled by ceiling height and unoccupied floor space with 1.5 sec being typical. But the cultural hall usually serves as a gymnasium, in addition to its other uses, and has durable wall sound-absorbing panels bringing the empty reverberation time down to 1.7 sec or less, meaning a full occupancy reverberation time of about 1 sec. Seating is folding chairs. The chapel, sound lock, and cultural hall are bracketed or surrounded by corridors, and usually there are classrooms and offices on the perimeter. In noisy locations, the corridor will be at the perimeter, and the core of the building will contain the classrooms and offices as well as the chapel, sound lock, and cultural hall. The Mormons have devoted considerable thought to the acoustical design of these buildings, while keeping the additional goal of economical construction in mind. Song and music, in general, is intrinsic to Mormon culture, and all local stakes and wards, as their local groups are named, will prefer to have custom-installed pipe organs. Their rapid expansion means that often electronics are a long-term temporary substitute, even if the buildings are always planned for the eventual installation of an appropriately sized pipe organ. The large Mormon Tabernacle organ, still including pipes fabricated by the original Utah settlers, is considered one of the world's great organs.

Acknowledgments

For helpful discussions and contributions, we gratefully acknowledge Dan Clayton (Clayton Acoustics Group, Carmel, NY); Dennis Fleisher (MuSonics, Grand Rapids, MI); Paul Henderson (Design2020, Harrisburg, NC); Rev. Gregory J. Hoppough, C.S.S. (Blessed John XXIII National Seminary, Weston, MA); David May (DCi Sound, Marcellus, NY); Francesco Martellotta (Dipartimento di Architettura e Urbanistica, Politecnico di Bari, Italy); Ben Markham (Acentech, Cambridge, MA); Courtney McGinnes (Acoustic Dimensions, Dallas, TX); James Mobley (Renkus-Heinz, Foothill Ranch, CA); Gerrit Duran (Duran Audio, Zaltbommel, The Netherlands); Peter Stoltzfus Berton (All Saints Church, Worcester, MA); Glen Sloan (Johns Creek Baptist Church, Alpharetta, GA); Peter M. Naranjo (Roman Catholic Diocese of Springfield, MA); Jason Summers; Richard W. Lee; Fr. Anthony Ruff, OSB (Saint John's University and School of Theology-Seminary, Collegeville, MN); Eric Wolfram (Riedel & Associates, Milwaukee, WI).

Mosques

M.1 King, G. R. D., *The Historical Mosques of Saudi Arabia*, Longman Group, United Kingdom (1986).

M.2 Abdou, A. A., "The measurement of the acoustics of mosques in Saudi Arabia." *Journal of the Acoustical Society of America*, vol. 113:3 (2003), pp. 1505–1517.

M.3 Orfali, W., "Room Acoustic and Modern Electro-Acoustics Sound System Design during Constructing and Reconstructing Mosques." Dissertation, Technical University of Berlin, Germany (2007).

M.4 Hasson, R., *The Mood of the 'ud*, Museum of Islamic Art, Jerusalem, Israel (2003).

M.5 Semi-permanent exhibition of Islamic Architecture at the Museum of Islamic Art, Jerusalem.

M.6 Visits by authors to mosques.

M.7 Khalidi, O., "Mosques in the United States of America and Canada." U.S. Government Printing Office, U.S. Embassy. Germany (2006)

M.8 Abdou, A. A., "Comparison of the Acoustical Performance of Mosque Geometry Using Computer Model Studies. Eighth International IBPSA Conference, Eindhoven, The Netherlands (2003).

M.9 Abdelazeez, M. K.. et al., "Acoustics of King Abdullah Mosque." *Journal of the Acoustical Society of America*, vol. 90:3 (1991), pp. 1441–1445.

M.10 Khalidi, O., "Import, Adapt, Innovate-Mosque Design in the United States." *Saudi Aramco World* (November/December 2001).

M.11 Karabiber, Z., "Acoustical Problems in Mosques: a Case Study on the Three Mosques in Istanbul." *Journal of the Acoustical Society of America*, vol. 105:2 (1991) pp. 1044.

An easily readable introduction to the Muslim faith can be found in:

Saeed, A., *Australian Muslims—Their Beliefs, Practices and Institutions.* Commonwealth of Australia (2004). ISBN: 0-9756064-1-7.

Acknowledgments

Thanks are due to Imam Muhammid Osman who introduced one of the authors to the essentials of Islam and proved the essential unity of faith of Islam and Judaism. Born and educated in Egypt, for many years he served as imam of Manhattan's Islamic Center and oversaw its move from a converted Riverside Drive residence to a magnificent new Mosque at Third Avenue and 96th Street. There, the Islamic Center welcomes people of all faiths to visit during the regular Friday services. Modest dress is required, and women have a separate praying room and can visit the main area outside of prayer times. Shoes must be removed before entering the praying area, and the visitor should attempt to coordinate body positions with those of the congregation to the extent practical.

INDEX

G

H

T

U

V

W

Y

Z